HIRING THE BEST

KNOWLEDGE WORKERS, TECHIES & NERDS

◆ ◆ ◆

THE SECRETS & SCIENCE OF HIRING TECHNICAL PEOPLE

by JOHANNA ROTHMAN

foreword by GERALD M. WEINBERG

DORSET HOUSE PUBLISHING
353 WEST 12TH STREET
NEW YORK, NEW YORK 10014

Library of Congress Cataloging-in-Publication Data

Rothman, Johanna.
  Hiring the best knowledge workers, techies & nerds : the secrets & science of hiring technical
people / by Johanna Rothman ; foreword by Gerald M. Weinberg.
      p. cm.
  Includes bibliographical references and index.
  ISBN 0-932633-59-5
  1. High technology industries--Employees--Selection and appointment. 2. High technology
industries--Employees--Recruiting. 3. High technology industries--Personnel management.
I. Title.
  HF5549.5.S38R68 2004
  658.3'11--dc22   2004016872

Trademark credits: All trade and product names are either trademarks, registered trademarks, or service marks of their respective companies, and are the property of their respective holders and should be treated as such.

This publication is designed to provide accurate and authoritative information with regard to the subject matter covered. It is sold with the understanding that neither the publisher not the author is engaged in rendering legal or other professional advice. If such assistance is required, the services of a qualified professional should be sought.

Cover Design: Nuno Andrade
Executive Editor: Wendy Eakin
Senior Editor: David W. McClintock
Editor: Nuno Andrade
Assistant Editor: Vincent Au

Distributed in the English language in Singapore, the Philippines, and Southeast Asia by Alkem Company (S) Pte., Ltd., Singapore; in the English language in India, Bangladesh, Sri Lanka, Nepal, and Mauritius by Prism Books Pvt., Ltd., Bangalore, India; and in the English language in Japan by Toppan Co., Ltd., Tokyo, Japan.

Printed in the United States of America

Library of Congress Catalog Number: 2004016872

ISBN: 0-932633-59-5   12   11   10   9   8   7   6   5   4   3   2

Dedication and Acknowledgments

I dedicate this book to my family—Mark, Shaina, and Naomi—for whom I shall be forever thankful.

I am grateful as well to many friends and colleagues for their review of early drafts of my manuscript. Any mistakes that remain are mine! In addition to the editorial staff at Dorset House Publishing, I thank Nicole Bianco, Esther Derby, Dale Emery, Paul English, Sally Hehir, Erik Hemdal, Elisabeth Hendrickson, Cem Kaner, Bob Lee, Vijay Manwani, Jonathan Ostrowsky, Janna Patee, Bret Pettichord, Dwayne Phillips, Barbara Purchia, Rob Purser, Steven Robbins, and Sally Silver for your varied and invaluable comments—each of you have contributed in ways I could never have expected. I thank you all.

Contents

Illustrations

Foreword

I've been consulting with high-tech firms for half a century, and I sure wish
I had this book fifty years ago. I cannot even estimate the number of times
I've seen hiring problems that would have been prevented by a manager
reading *Hiring the Best Knowledge Workers, Techies & Nerds.* At least once in
every consulting assignment, a manager asks me a question that could be
answered by a quick look-up in Johanna Rothman's contents or index. I
know, because many times in the past decade, I've recommended Johanna
as a hiring consultant to my own clients, and she's never failed to produce
phenomenal results.

Hiring mistakes cost high-tech organizations literally billions of dollars
each year, plus untold pain and anguish on the part of hiring managers,
applicants, and employees. This superbly organized book distills
Johanna's many years of experience doing just what its title implies—
Hiring the Best. It nicely balances cases drawn from that experience with
principles abstracted from dozens or hundreds of cases. Indeed, it would
be a poor manager who couldn't pay for his or her own salary by applying
these principles to an organization's hiring process.

There's only one thing I can find wrong with this great book: The
author underestimates its value. She wrote it as a book for managers who
are hiring, but that's too narrow an audience. It excludes several other
large groups whose members should be reading it:

1. Everyone who participates in the hiring process, such as
   coworkers who are called upon to interview job candidates
2. Teachers and trainers who prepare students for jobs and
   need to understand the processes that will be, or should be,
   used to select them

3.   Any knowledge worker, techie, or nerd who is now, or will be in the future, applying for a job, or even for a promotion in his or her present job

But, in the end, *Hiring the Best Knowledge Workers, Techies & Nerds* is a book for managers, and any high-tech manager who doesn't read it as a hiring manager may soon be reading it as an out-of-work applicant looking for a new job.

July 2004   Gerald M. Weinberg
Albuquerque, New Mexico   Author of *The Secrets of Consulting*
and *The Psychology of Computer Programming*

Preface

I've had the opportunity to hire or participate in the hiring of hundreds of technical people over the years, including developers, testers, technical writers, technical support staff, pre- and post-sales applications engineers, consultants, leads, and their managers. I've been part of interview teams charged with hiring product managers, electrical engineers, mechanical engineers, in-house teaching staff, and information systems staff.

Hiring technical people has never been easy. Many organizations persist in a near-constant state of having too few *qualified* technical workers. When the economy is strong, we attribute the shortage to too few qualified candidates to fill the many openings. When the economy is weak, we attribute the shortage to too many poorly trained applicants with unsuitable backgrounds.

Too often, we apply the same hiring techniques to knowledge workers that we use to hire skill-based staff. Skill-based staff members possess a set of tools and techniques that can be applied in the same way in almost all situations. Technical people—in particular, knowledge workers—must adapt their knowledge to the specific situation. Such workers are not just the sum of their technical knowledge; they are the sum of both *what they know* and *how they apply that knowledge* to the product. In particular, how they use their technical skills to benefit the product, how they manage their work, and how they manage their relationships with other people all must be assessed when hiring and evaluating a knowledge worker.

While there are some similarities in the hiring process, hiring technical people—knowledge workers—is vastly different from hiring purely skill-based staff.

Knowledge workers have unique qualities, preferences, and skills—such workers are not fungible assets.[1] The ability to adapt knowledge and to innovate makes one developer, tester, project manager, or technical manager different from another. That difference among people is key to making good hiring decisions.

You want your organization to succeed, and so you need to know how to define and assess a technical candidate's qualities, preferences, and skills, but you also need to be able to predict a technical person's chance at succeeding in *your organization*. The techniques and recommendations set forth in this book are designed to make hiring a streamlined, efficient, and satisfying experience.

Why read this book?

"The whole interviewing thing takes forever."

"How do I know this candidate will work out?"

"I can't seem to find candidates who meet the job's specifications."

I hear comments like the preceding every day. The comments express the frustration that many technical managers feel as they attempt the very difficult job of hiring. If you're like most of the technical managers I've worked with, you may not be sure how to define the job's requirements, how to find suitable candidates, what skills you need to interview well, or how to make an offer that the candidate will accept. Most technical managers who ask me for guidance in hiring technical staff find it difficult to define appropriate requirements, to assess experience and cultural fit, and to check references in a way that makes sense for the candidate and themselves.

Or, possibly, you know how to do all of that, but the hiring process consumes more of your time than you comfortably can allocate. If you have any of these problems, the material I present in this book can help you.

Many books on hiring give good advice on how to ask questions in an interview or help you develop reference checklists. But few books address how a technical hiring manager can create an efficient and effective hiring process for a technical organization. If you hire technical people, you know that the approach used to interview, assess, and hire skill-based staff does not do the job when you're trying to hire knowledge workers or to evaluate the skills of those technical people already in your group. You need to precisely define the specific experience, qualities, preferences, and

---

[1] For a detailed exploration of "fungible assets," see Tom DeMarco, *Slack: Getting Past Burnout, Busywork, and the Myth of Total Efficiency* (New York: Broadway Books, 2001), pp. 13-21.

skills you want people to have, and you'll need a specific strategy to help you detect whether a particular candidate has the necessary expertise.

Here's what I hope this book can do for every hiring manager who uses it as I've intended it:

- Save you time and money every time you hire.
- Help you hire people who can perform the required work well.
- Help you screen, evaluate, and hire the right staff for your specific organization.
- Eliminate the wasted time and suffering that result from having to fire people who should not have been hired in the first place.
- Help you develop and demonstrate fundamental management competency.[2]

Save time and money. This book offers a streamlined approach to hiring. The more you streamline your hiring approach, the faster you will be able to evaluate suitable candidates, and the better the hiring decisions you'll make. An effective hiring process is especially important when you consider the toll a bad hire can take on your organization. Add up the direct monetary costs of recruiting, the cost of the time you and your staff spend on hiring the person, and the actual cost of the person's salary and benefits while he or she works for you, and you'll quickly see that the cost of a bad hire can be enormous. Perhaps equally costly, a bad hire saps energy from the work your organization is trying to perform, and prevents the work from moving forward. A bad hire doing substandard work can even damage your product to such an extent that it will have to be redeveloped from scratch.

Hire people who can perform the work. Few things are worse than feeling that a new employee somehow misled you on his or her résumé or in the interview. You thought you hired Dr. Jekyll, but Mr. Hyde showed up to work. Or, you thought you hired "the best and the brightest," but the people who came to work seem mediocre and dim. If you use the job analysis template in this book to help you define the qualities, preferences, and skills needed for a particular job, you can create a set of strong interview procedures that should prevent you from simply hiring just a warm body.

---

[2] For a solid treatment of the hiring manager's role in finding talented staff members to hire, see Ed Michaels, Helen Handfield-Jones, and Beth Axelrod, *The War for Talent* (Boston: Harvard Business School Press, 2001).

Screen, evaluate, and hire the right staff for your specific organization. I've learned from my early hiring mistakes and know ways to avoid making those mistakes again. There *are* good people out there, no matter the state of the economy. You can help yourself hire well by first defining a standard for what "a good employee" is for your specific organization, and then translating that standard into precise job requirements, a sound job description, and a comprehensive listing of information needed for successful interviewing.

Fire fewer people. Most managers dislike the firing process: the warnings, the get-well plan, the actual firing. Many ignore the problem altogether or shunt the non-performing employee to other projects or other managers. If you would like to avoid the firing problem, use this book to help find people who can perform the work.

Develop and demonstrate your own management competence. If you'd like to be a better manager or you'd like to advance in management, this book can help you hire well. Simply put, managers who can tell the Jekylls from the Hydes will be more successful than managers who cannot. Likewise, a manager with a staff whose members can work with others in the company will be able to complete his or her assignments faster and with greater success.

The bottom line: Once you've defined what a "good employee" means for your needs and your culture, you can quickly review résumés, conduct interviews, make offers, and hire the right technical person for the job. I hope the numerous tips, suggestions, and recommendations in this book will help you expedite the hiring process as well as make it a more pleasurable experience.

June 2004   J.R.
Arlington, Massachusetts

HIRING THE BEST

KNOWLEDGE WORKERS, TECHIES & NERDS

Part 1

Defining Requirements for Yourself and Your Candidates

The three chapters in Part 1 deal with defining requirements: developing the hiring strategy, analyzing the job, and writing the job description.

The hiring process can be viewed as a series of steps, each of which requires some preparation. The schedule I recommend you use for successful hiring follows, showing how much time I generally allocate for each step.

From beginning to end, the total time per candidate should be approximately one day plus the time you'll spend on sourcing activities and résumé review. If you're spending more than one day on each candidate, review your preparation work. Don't waste time on things you can pre-plan or organize. Define the job so everyone who's recruiting on your behalf understands the open position. In addition, if you're spending more than a few weeks recruiting candidates, reorganize your recruiting mechanisms to recruit more effectively.

Preparation (Time: 2 hours per open job)

Step 1. Define the requirements for the job:

a. Define your hiring strategy, identifying why you're looking for people (30 minutes).
b. Analyze the job, detailing the requirements a person should satisfy to be successful (30 minutes).
c. Write the job description, describing what interviewers should look for when evaluating candidates (30 minutes).

Step 2. Write and place the job advertisement (30 minutes).

3

Sourcing (Time: 3 hours per candidate)

> *Step 1.* Select your sourcing mechanism—that is, the techniques you'll use to attract suitable candidates. Work with your HR staff to implement those techniques (10 minutes, plus actual recruiting time if you attend job fairs or other networking events).
>
> *Step 2.* Filter résumés, reviewing each to determine whether you want to phone-screen the candidate (30 seconds per résumé).

Interviewing (Time: 3 hours per candidate)

> *Step 1.* Define the list of questions you'll ask each candidate on the phone to qualify him or her for an in-person interview (10 minutes).
>
> *Step 2.* Phone-screen each candidate, conducting a brief phone interview to determine whether you want to interview the candidate in person (10-45 minutes).
>
> *Step 3.* Schedule the in-person interviews, selecting a team of interviewers and planning who will ask which questions when (60 minutes).
>
> *Step 4.* Hold a follow-up meeting with interview-team members to hear their perspective on the candidate (15 minutes per candidate).

Offers (Time: 2 hours per offer)

> *Step 1.* Check references (60 minutes).
>
> *Step 2.* Extend the offer (60 minutes).

Before you read further, I must issue my own requirement: Some of the advice in this book may conflict with your organization's hiring policies and practices. If at any time you're not sure whether something I've suggested is appropriate for your organization, your corporate culture, or even your geographic location, check with professionals in your company's Human Resources or Personnel Department, or with your corporate lawyer. They will factor in conditions specific to your particular situation; above all else, *follow their advice.*

1

Developing Your Hiring Strategy

"I wish I'd never hired Zeus. I know he's excellent at his technical work, but he's so difficult. He intimidates people, hurling words when he's displeased. What a mess."

—Statement from a dissatisfied manager

Hiring "messes" do happen—but they are something we can and must work to avoid. In truth, the decision to hire an employee is one of the most critical decisions a manager can make. You, your team, and the organization will live with the long-term consequences of your hiring decision. Getting it right the first time can be challenging but, with a good bit of hard work and intelligent planning, it can be done. When you invest your time developing a hiring strategy that defines the kinds of people you need and that helps you determine how such people will fit into your organization, you can improve your success ratio dramatically, perhaps even to the point of achieving a perfect record. A well-constructed strategy should pay big dividends for the future of your group.[1] One good way to start is by asking questions such as the following:

1. What kinds of people are you looking for?
2. Which roles do you want to fill first?
3. What talents or skills are most important to your search?
4. How will you decide on a candidate?
5. What will you do if you can't find the right people?
6. How do you know when you've got the right person?

---

[1] For guidance on linking your corporate strategy to hiring, see Ed Michaels, Helen Handfield-Jones, and Beth Axelrod, *The War for Talent* (Boston: Harvard Business School Press, 2001).

To help find answers to these questions, let's look at some real-world scenarios, tales taken from the workplace, possibly even from a workplace you'll recognize.

Kent, a test manager, has a great test group. He needs to hire someone else, and his first thought is, "I need someone just like Louise; she found all those bugs in the last release." *Should Kent be looking for other testing skills?*

Sharon, a development manager, suspects that her group's architecture skills are weak, but because the people in her group all work well together, she doesn't know what to do. *Should she hire the senior architect who might not get along well with the rest of the group, or someone who can fit into the group well? Does she have to choose between the two options?*

Dan, a hiring manager, just fired his fourth new hire in three years—an unusually high percentage of the ten people he's hired in the same time period. Even though this most recent ex-employee's technical skills were appropriate to the job, the employee consistently infuriated and alienated his coworkers. *Is Dan hiring the best-suited people for the culture?*

Beth, a technical support manager, works for a company that is changing its product line from an expert-use-only product to a suite of products that will be used by a range of people with various levels of expertise. Her staff is great at supporting Ph.D.s, but can its members support administrators and lab technicians also? *Should Beth be hiring from a different skill-set?*

Each of the four tales depicts a common hiring problem, possibly one you've experienced firsthand, but I'm not going to tell you what I think Kent or Sharon or Dan or Beth should do. Instead, I'll arm you with the questions I asked them to consider, questions you, too, can use to develop your hiring strategy:

- *Should you hire someone just like yourself or just like the rest of your staff?* Sometimes you may want to hire someone who will fit easily into the team. At other times, you may want to hire someone completely different, so that you can take advantage of differences in experience, skills, and personal qualities.
- *What are your tradeoffs in your hiring?* Do you have only one requisition authorizing you to hire one candidate, but you need two people? If you have to make tradeoffs, you'll not only have to decide which aspects of the job you're willing to sacrifice, you'll also want to make sure that the candidate can perform the most important work successfully.

- *Are you looking either to overcome certain weaknesses or to expand certain skills in your group?* Kent, our test manager, may decide to bring in someone with a wide range of experience using automated testing tools in order to try different kinds of testing for the product. Or he may decide that another exploratory tester has just the right mix of skills.
- *Are you hiring based on a candidate's specific technical skills, but firing people because they can't fit into your culture?* If you find yourself having to fire people within a year of their hiring, the reason may be that you have trouble assessing how a candidate will fit into your group or into the larger organization.
- *What kind of diversity are you trying to achieve in your group?* Having only senior-level developers or manual testers in a group assures that the group will lack skill diversity. Staffing with only men or only women indicates another lack of diversity. Having only extroverts or only introverts is yet another form. Maybe the members of your group know everything there is to know about using the development tools but not how the customers actually use the system. When you consider diversity, don't just look at the obvious; look at *all* the issues associated with skill level, experience, and how people communicate.[2]
- *Are you looking for a personality who will complement and expand the capabilities of the current team, or are you looking for someone who will fit into the team seamlessly?* If you have a group that already solves problems well as a unit, maybe you should hire someone who fits into that group. However, even a group whose members work well together can use some jiggling every so often, if only to spark new ideas.

A well-thought-out hiring strategy will help you identify the candidates you should interview.

Ask questions when creating a hiring strategy.

Before you determine your hiring strategy, there are several important questions to consider:

---

[2] Diversity can be enhanced through your hiring strategy. For sage advice on other factors to consider than just a candidate's qualifications, see Ed Michaels, Helen Handfield-Jones, and Beth Axelrod, *The War for Talent* (Boston: Harvard Business School Press, 2001), pp. 13ff.

- What problem(s) are you trying to address with the hiring you're planning, and do you have the facts you need to assess your current team's qualities, preferences, and skills?
- Which roles are most important to fill, and in what sequence?
- What are your criteria for choosing suitable candidates?
- What is your decision-making process?
- What contingency or risk-mitigation plan will you use if you can't find the right people to hire?

Let's start by considering the problem you want to solve with your hiring.

Identify the problems you should address.

Some hiring managers begin the hiring process by compiling a wish list of technical skills (two years of C++, four years of Java, five years of Unix, three years of project management, and so on). Unfortunately, a detailed skill listing tends to constrain the position, so that no candidate perfectly fits the role.

Instead of focusing on technical skills, think of hiring as a problem-solving exercise. I focus on five problem-solving steps:

- Define the problem you hope to solve by hiring this person or these people.
- Develop a strategy for identifying the candidates who are best-suited to your needs.
- Assess your current staff to see where you need complementary skills or experience.
- Define the kinds of people you require.
- Hire the person or people you need to solve the problem.

In the following paragraphs, I describe common problems and some ways to handle them. Some problems may resemble ones you have been trying to solve.

You need additional people to do more of the same kinds of projects that are currently being done. If you need to hire more people to do the same kinds of projects, put extra emphasis on each candidate's technical skills, *if you can find enough candidates*. If you can't find technically skilled candidates, then look for candidates who fit into the existing culture or who have demonstrated an ability to adapt and learn—and plan to train your new hires. Consider whether you need junior-level people, senior-level people, or people who are experienced technical leaders. Especially when you're hiring a lot of people at one time, make sure you don't base your decision to eliminate candidates solely on the fact that they seem to have too little or too much experience. Junior-level candidates can grow along with your organization, and can be the leaders in a few years. Senior-level candidates can bring significant problem-solving expertise into your organization.

Your work is changing focus from one kind of work to another. If project staff members must make the transition to a different kind of work, you may need to add people who are different from those currently on your staff. One test manager recently told me, "My folks are great at testing the product from a black-box perspective. However, that's all we do. With this new product, I need to modify the testing to include performance and reliability testing, something people on my staff know nothing about. They just don't have the technical background to know how to perform this kind of testing." This problem is especially challenging if you cannot add staff, but must lose current staff to make room for people who are qualified to perform the new assignments. In this case, you'll want to pay special attention to the required functional skills for the job.

Your technology is changing from one technology to another. When you're recruiting because your company must make the transition to another technology, consider a candidate's problem-solving skills, adaptability, and cultural fit for the new organization, rather than focus on the person's current technical skills, particularly since you'll need to train your staff in new skills anyway. For example, if staff members in the new technology use a different programming language from the language used in the old environment, it's easier to assess suitability among candidates who've already learned multiple languages, rather than selecting people who have worked in only one language.

Decide how many people you need, at both junior and senior levels, and in which levels new staff need to possess technical expertise. One option to consider when moving to a new technology is whether to hire someone with significant expertise in the new technology to mentor both your current staff and new hires. Keep in mind that in order for this technical mentoring to be successful, the expert will need to build a rapport with the team quickly.

If you're adding new technology and still supporting the old technology, don't hire a new team to work on the new products and keep the existing staff working on the old products. The existing team may want to work with the new technology and may become frustrated that new people will have all the fun work. If you break the work up into new (read, "exciting") and old (read, "boring," "tech. support," "house-keeping") work, you'll create more problems than you solve. If you want to retain your current staff members, ask them what work they want, and hire to backfill their current roles so they can move on to do the new work.

You're on the cutting edge of technology but you're so far ahead of the market that you don't know what technical skills the staff should have. Sometimes, when you're on the cutting edge, you may find it difficult to determine precisely which technical skills will be needed. Here, a good strategy is to place emphasis on a candidate's adaptability, cultural fit, and ability to work in teams. Consider the experience level and the technical leadership abilities of the candidates. Think in terms of what work, call it "X," must be done in your cutting-edge project. That way, even if you don't know the specific required skills, you can ask candidates to describe their experience doing "X."

Years ago, before configuration management systems were common, I needed to hire a release engineer, someone with expertise in builds, branches, and what we now call configuration management. Since I didn't know precisely what skills would be required, I looked for a candidate who could communicate well with system developers, and who had demonstrated an ability to organize complicated work and run smoke tests. I suspected I didn't need someone with years of experience, just an exceptionally good problem-solver. The candidate I hired had only two years of experience, but he'd worked as a programmer throughout high school and college. He was a great release engineer, and now is a highly qualified configuration manager.

You're putting together a brand-new group or are adding staff to a recently established group, but you don't know what personalities, characteristics, and skills will make the best mix. If you have a newly formed group or are adding people to a group that has not been together very long, the people you add should enhance the group's ability to work together and mesh; they should not prevent group members from working well together. Hiring a personality who doesn't fit well will prevent your team from doing the work. An established group, whose members are confident of their abilities, can handle different personalities and challenges to the current work; new groups are less likely to succeed.

In order to be successful, a new group needs to build confidence and develop ways for individual members to work together. You'll want the most experienced people you can afford, because you need people who can manage their work while developing healthy, working relationships with coworkers. For this group, you need people who are experienced in both technical and communications skills. As your group matures, you can hire less experienced staff.

You need to change what your group can deliver, but you don't know what skills to add. Maybe you have a great group of developers and it's time to add some testers. Maybe you've got writers, and you need some editors. Maybe you've got manual testers, and it's time to add some automation to the mix. Whatever the case, if you're looking for people to fill a gap, you'll want to consider functional skills, but don't forget to assess each candidate's cultural fit. Because new people already will be rocking the boat just by virtue of performing a new function, they particularly need to fit in with the team and the culture.

Adding people with different skills to your team tests its maturity and adaptability. Your challenge is to overcome the second-round effect in which new people join an established team but are not perceived as full partners. Look for people who, in addition to possessing superb technical skills, can quickly learn and adapt to the team's culture.

I once worked with a manager who brought ten new people into what had been a four-person group. He had hired the new members on the basis of specific skills (user-interface development, testing, and so on), but he did not consider how well they would fit into the existing team. After sixteen months, it was obvious, even to an outsider, who the original members were and who'd been hired later. The fourteen-member team couldn't make the project succeed until both the original team and the new members changed their behavior and adapted to each other. If the manager either had hired more adaptable people for the original group or had focused his second-round hiring on people with better communications skills, the team would have meshed much sooner.

Your group must finish a project faster than originally planned, but you don't know whether adding staff will increase productivity. The good news is that you've got a group of people who work well together, but you need to increase productivity. Sometimes, adding people to a team is the answer to attaining a faster release, but bringing them up to speed may counteract the contribution that additional staff should eventually make. If productivity can be increased by assigning new people to work on parallel projects or if new members can work in parallel with the original staff on one project, and management can handle such a challenge, by all means add staff. If you make the decision to add staff, bringing in candidates who fit your culture is critical.

You need a few additional people right now, but they won't be needed forever. Sometimes, you need people *right now* on a project, but you don't want to keep them in the company long-term. If this situation is likely to arise, you might choose to hire fewer permanent staff members and more contractors. Analyze your immediate, mid-term, and long-term needs to decide which of your candidates should be offered a contract and which should be offered employment.

When you hire contractors, consider their communications skills to be equally as important as their technical skills. If you don't intend to keep someone around for a long time, you'll need that person to be conscientious about thoroughly documenting what he or she does, detailing how all aspects of the job are performed. Make sure that when the contract term expires, the contractor is capable of handing off the work to other people. For the most part, I interview contractors the same way I interview permanent staff, although I do give greater emphasis to such areas as their ability to complete and hand off work, and their communications skills. I treat this important topic in greater detail in Chapter 7, "Developing Interview Questions and Techniques."

You have to fire more of the people you hire than seems reasonable, but you aren't sure how to get a more stable staff. If you find that many of your new employees are not successful at their jobs, or that you fire even 5 percent of your new hires, reassess both the content of your interview questions and how you or your interviewers ask those questions. The most effective screening involves behavior-description questions that include some combination of technical-skill and cultural-fit analysis. Do you and your interviewers have enough technical and interviewing expertise to ask the appropriate technical questions and assess the answers? If not, you'll need to change interviewers, and increase the interview team's level of expertise.

If you have to repeatedly fire people because their technical expertise is inadequate, you're probably not asking specific-enough interview questions. If you have to fire people because they don't fit into your group, perhaps you haven't fully identified the kinds of people that best fit your culture.[3] If you have thoroughly defined the kinds of people you need and you still must fire people because they don't fit into the organization, you're probably not using a consensus-based approach to candidate appraisal. You'll discover more about the candidate's qualifications and his or her fit with your culture if you invite several people outside of your

---

[3] For a sound discussion of hiring and firing complexities, see Jim Harris and Joan Brannick, *Finding & Keeping Great Employees* (New York: AMA Publications, 1999), pp. 18ff.

group to be part of your interview team. Generally, when you use a consensus-based approach to candidate appraisals, you develop more of an understanding about what your team wants. This topic will be treated in more depth in Chapter 10, "Following Up After the Interview."

A True Story

Fred is a non-technical, quick-to-judge MIS manager. When I first talked to him, he boasted that he could interview someone and know within thirty seconds whether the person would fit into the organization. I didn't hear from Fred again until after he'd fired two people before their three-month anniversary with the company. More than a little bit rattled, Fred decided to ask for my help with interviewing and the hiring process. I suggested that Fred recruit people from the rest of his company to help him with the interviewing, and gave him guidelines to follow. He assembled a group fairly quickly that included an MIS technical staff member, the release engineer, the support manager, and a couple of developers—all people who understood the implications of MIS work and who possessed some of the expertise required for the jobs to be filled. Once assembled, the interview team posed questions they wanted to ask, and Fred agreed to let them go ahead with the interviews. Fred also agreed to withhold rushing to judgment during each interview. Following each interview, the interview team met to discuss the qualifications and suitability of each candidate. Happily, this approach enabled the team to find two candidates to replace the fired employees, producing an MIS group that remains stable and successful to this day. At last check, the new employees were still working at the company, almost three years later.

You have too much turnover, but you don't know how to reverse the trend. If your employees choose to leave after they've worked at your company for only a year or two, maybe they were not the right candidates to hire in the first place. Unless you have defined the job as an entry-level, short-term position, you do not want people to view your company as a temporary port in a storm. Such employees are not good long-term investments and are not worth the training you'll undoubtedly need to invest in them. If you find yourself facing excessive turnover, you're probably not asking the right interview questions.

A test manager stated that she'd replaced five members of her ten-person group in one six-month period. As this turnover rate was unusually high, I recommended that she contact the former employees to retroactively conduct exit interviews in order to find out why so many people had left. The exit interviews gave her the reason: She learned she had consistently hired people who were risk-takers who enjoyed solving problems in unique ways. The development organization didn't value those testers, and wanted to work with testers who planned testing in a predictable way. The hiring problem wasn't that she was hiring people with poor technical skills; the problem was that she hadn't given enough thought to the cultural-fit problem: how to hire risk-taking testers who had enough patience to continue working through the changes she was trying to implement. By changing her cultural-fit questions to identify how the testers tested, and by looking for people with patience for cultural change, she was able to keep turnover to a minimum.

Your ability to recruit more people to join an existing project has become increasingly difficult, but nothing you do seems to change the picture. If finding people is difficult, maybe you're not using enough different approaches or recruiting mechanisms. If you use only one recruiting method, you run the risk of missing out on potential candidates. For example, if you only use classified ads, you'll miss people who only work with recruiters. If you only use one general-purpose, Web-based job board, you'll miss people who use industry-specific sites or geographic sites.

For more details on how to build and use your recruiting network, see Chapter 4, "Sourcing Candidates." Also, check to see that you're not inadvertently discriminating against people who are different from you. Chapter 6, "Reviewing Résumés," describes ways to check for your prejudices.

You have a solid group of people on the project, but everyone has the same set of skills or philosophy of work. You'd like to add more diversity. Sometimes, when a group has been together a long time, its members may start to think alike. The best remedy for this is to shake up the group by gradually adding people with different personality types or backgrounds. If you're changing the focus of your product base, you might add people who more closely reflect your customer base. Or, if the team consists of people primarily of one gender, race, or philosophical outlook, hiring people of the other gender, from another race, or committed to other philosophies will make for a richer work environment.

In technical groups, technical skill and expertise are usually valued more highly than personality or race or gender. That's not to say there are no bigots or prejudiced individuals in technical groups, but that, in my experience, most people are more interested in what another person can do than in what the person looks like or how forceful his or her personality is. Not surprisingly, we tend to neglect considering personality diversity while hiring. How a person solves a problem or performs an assignment is influenced by his or her personality, and it can be used to advantage in matching a candidate to a job. For example, many people working in the technical field are quick to make decisions, but creative product architects may choose to ponder several designs, looking at the pros and cons before coming to a conclusion. Some testers like to plan their work; others explore a less structured path as opportunities arise. Some people prefer to talk out the issues; others prefer to think about the issues privately and then discuss them.

Look at the range of personalities on the existing team to see whether all team members have one kind of personality. The more diversity you have in personality types, the less likely you are to be blind-sided by a problem no one considered.

Sometimes, diversity can be achieved by mixing experienced workers with entry- or junior-level staff members. Such a mix would have bene-fited one development manager who told me, "Everyone in my group has at least ten years of valuable experience. Most are designers, but we also have three real architects. Unfortunately, I don't have enough high-level work to keep them all busy right now. I need junior people to be my senior folks' journeymen, so I have a more natural mix of engineers."

You can hire junior-level people to perform jobs that do not require senior-level knowledge and talent. Allow for the maturity hierarchy of technology skill and knowledge to take a natural path—mix experience and knowledge levels.

You need more management capability, but no one you've inter-viewed is qualified. If your group has grown in size, or if you have a start-up group that must make the transition to the next level to become a more productive entity, you may need to hire more managers. The chief technology officer of a Web-based start-up typically might manage a tech-nical group of twenty software developers, testers, and operations staff members for years with only the help of technical leads in the various functional areas, but if the organization decides to hire another five people, then experienced, full-time managers, not just technical leads, will be needed to make sure all of the management tasks will be accomplished correctly, on time, and within budget. In Chapter 14, "Hiring Technical Managers," I provide detailed information about hiring managers.

Sometimes—most times, in fact—you have *more than one problem to solve.* When that happens, list all your problems in order of their priority so you can then determine whether you need to hire additional staff members based on technical skills, cultural fit, adaptability, technical leadership, or some other quality, preference, or skill. Once you've defined each problem and determined its importance in relation to the other problems that also must be solved, you can choose which types of people you need to hire first. This topic is discussed more fully in the following section.

Determine which roles you want to fill first.

When you're hiring more than one person, or hiring into a group over time, decide which capabilities are your highest priorities. Not all roles in your organization are the same. If you need a product architect, then a designer will not do. Whether you're hiring for a product company or for an information services group, consider the kinds of people you want and your priorities.

You will need to make decisions as you build your list of first, second, third, and so on, hires. Some typical tasks are identified in Table 1-1, below, which suggests job titles to fill function areas:

Function to Be Performed	Possible Roles and Job Titles
Requirements Analysis	systems engineer, analyst, requirements specialist
Development	systems architect, senior designer, junior designer, programmer, project manager, technical lead
Release Engineering	build engineer, librarian, configuration manager, operations analyst
Testing and QA	automated tester, manual tester, exploratory tester, test project manager, technical lead, metrics gatherer
Documentation	editor, writer, book designer, technical lead, production specialist, graphics artist
Support	tier-1, -2, or -3 support (first-line, mid-level, and senior-level support staff)
Usability Engineering	interaction specialist, designer
Project Management	project administrator, project manager, program manager

Table 1-1: Function/Role Chart.

A True Story

A chief technical officer of a Web start-up defined his current hiring needs: "We've moved past the initial start-up phase. We have three developers—I guess I'd call them senior designers—and I've been doing the architecture. It's time to bring in a project manager and some testers, so we can 'product-ize' this beast now that we've got the funding. But, these people have to work *with* us, not against us. I'm not ready for formal release engineering, or formal process definition, or formal system tests, but I am ready to start automating tests of the product core. We need a technical project manager, an automation tester, and one more tester who can find the problems we developers don't see."

The CTO in the preceding story is trying to solve the staffing problem by filling in positions with other skills. Since his group is small, he's considering cultural fit (qualifications he says he is "not ready for"), but the driving force behind his hiring is to bring more people on-board to do different work than is done by the people he already has.

Once you've decided which roles you want to fill first, go back to assess your current staff members and the roles they perform, making sure they continue to fill those roles appropriately. If they still perform jobs that partially fill the problem areas you're trying to staff, include them on your interview team. When the people already fulfilling some of these roles participate on the interview team, you can obtain a richer picture of each candidate. If these current staff members can no longer perform the jobs you need done, determine how many of which kinds of new people to hire, and decide how you're going to manage the problem of your current staff's inability to perform the needed work. For more guidance, see Chapter 15, "Moving Forward."

If your hiring strategy includes hiring many people at one time, you may be lucky enough to find candidates for positions you need to fill but weren't planning on staffing until later. If this happens and you have the budget to support these additional employees now, hire them! Then, re-plan the work your group will do, and update your hiring strategy.

Decide which criteria matter most.

If one person could make all the hiring decisions, the hiring process could be accomplished comparatively quickly. However, in most organizations except possibly the smallest, one person probably can't ask all the questions and assess how each candidate would respond in different situations and to different people. To facilitate the process, create an interview team so that multiple people can interview candidates. Ideally, the same individuals you select for the interview team will be available to interview *all* the candidates. Once all members of the interview team have assessed each candidate's technical qualifications, proficiency level, cultural fit, and so on, invite interview-team members to compare notes on each candidate, encouraging them to share both positive reactions and negative impressions. Assure people on the interview team that their comments will be kept completely confidential. In some circumstances, it is advisable to require all members of the interview team to sign a simply worded confidentiality agreement prior to beginning discussion. Clearly, interview-team members must trust each other *and you* sufficiently to open up freely, but a confidentiality agreement can help to emphasize the seriousness of their involvement in the hiring process.

Train your interview team to apply a limited-consensus approach to hiring. When groups use limited consensus, not everyone may agree with the decision, but each person should be satisfied enough with a particular candidate's suitability not to block the decision to hire him or her. Limited-consensus discussion sessions provide the following benefits:

- Interview-team members feel valued because they are included in the decision-making process, which builds camaraderie and makes them more likely to help integrate the new staff member into the group.
- Interview-team members develop a stronger sense of their organization's culture, and learn how to successfully use it.
- The manager learns about the candidate from the different perspectives of each of the interview-team members.

The manager ultimately is still responsible for the hiring decision, but he or she doesn't have to gather and evaluate all of the data about each candidate single-handedly. For more discussion of how to appropriately use multiple interviewers, see Chapter 9, "Planning and Conducting the In-Person Interview," and Chapter 10, "Following Up After the Interview."

A True Story

> Steve, a development manager, described his interview team this way: "Our HR guy talks to each candidate to assess whether the person will fit into this small-company culture. Then, a senior developer asks difficult questions about the kinds of products this candidate has designed and how the candidate makes design decisions. The test lead then talks to the person about how he or she manages, develops, and tests code. One of the mid-level developers is great at developing audition questions, so he runs the audition part of the interview. The hiring manager comes at the end, following up with more problem-solving and cultural-fit questions to determine whether this candidate will fit into the group."

As Steve has described in the preceding True Story, everyone in the group has a role—and an audition-style format is integral to the process. Each interviewer takes a different aspect to investigate in each candidate. When Steve and his group members put all their observations together, sharing various perspectives, they have a clearer picture of the "whole" candidate. If someone sees a candidate as assertive during the interview, and another person sees that assertiveness as bossiness, the two interviewers can compare their different perceptions. For more on auditioning techniques, see Chapter 7, "Developing Interview Questions and Techniques."

Steve's group uses a limited-consensus, decision-making process. Since team members have all seen a different aspect of the candidate in the interview, they can now discuss the candidate and share their conclusions about the candidate.

A team of people is helpful for interviewing a candidate, and it makes sense to use that team both to evaluate the candidate and to help decide whether to hire the candidate.[4] With enough people on the interview team to contribute multiple perspectives, the hiring manager's job becomes easier. If you worry that too many perspectives may make it difficult or impossible to make a decision, read Chapter 10, "Following Up After the Interview." For more information about audition techniques, see Chapter 9, "Planning and Conducting the In-Person Interview."

---

[4] The topic of using a team for the process of hiring is nicely treated in Cem Kaner, James Bach, and Bret Pettichord, *Lessons Learned in Software Testing: A Context-Driven Approach* (New York: John Wiley & Sons, 2002), pp. 206ff.

Identify what process you'll use in decision-making.

Consensus-based hiring provides a great way to manage the potential risk associated with making a bad hiring decision. Here's what happened when one hiring manager, an SQA director, ignored the concerns of her group: "I desperately needed a release engineer. A friend of mine was available. I knew she could do the job, and I wanted to hire her outright. When she came in for the interview, half the people on my staff said she wasn't going to work out. I was desperate, so I hired her anyway, but as a short-term contractor. It's a good thing she was a contractor and not a permanent employee, because after she'd been here about seven weeks, she had made enemies out of almost everyone. People refused to talk to her and were on the verge of refusing to work with her. If she'd been a permanent employee, I would have had to fire her. The fact that she was a contractor allowed me to sever the work relationship without destroying our friendship. I will never ignore what my group says about a candidate again."

Decide who in your organization will help you make the hiring decisions. If a valued employee is voluntarily leaving, consider using that employee to evaluate potential candidates. I've had good results using such people to help interview their replacements.

Some managers worry about consensus-based hiring decisions, arguing, "Doesn't consensus-based hiring allow one employee to veto the entire process? What if the vetoer is someone I don't want to keep in the organization?" The answer to this is simple: In the interview process, only involve employees whose work you respect and value. If an employee isn't successful in his or her technical position, don't make that employee part of the interview team.

When you have an interviewer who consistently vetoes candidates, do the following:

1. *Make sure everyone realizes why you're hiring another person.* Determine whether anyone has anything to fear from this new hire. Deal with that fear now, before you begin the interview process. Otherwise, you won't be able to hire anyone.

2. *Clarify the job description.* Sometimes, people veto candidates because they don't understand the role you want filled in your organization. For more both on defining the problem you're solving and on clarifying the job description for inter-

viewers, see Chapter 9, "Planning and Conducting the In-
Person Interview."

3.   *When someone vetoes, ask why.* Drill down to discover the reasons, and check those reasons with other interviewers.

I learned of the following stalemate from a development manager in a
start-up organization: "As development manager, I tried to hire a second
developer into the group. The first interview day, we interviewed three
people. My on-staff developer's first reaction was, 'Not one of these guys
is any good. Time to look for more people.' I was surprised, and said so. I
asked why he thought none of the candidates was any good, because I'd
thought two were great, and one was okay. He said, 'They'll never make it
here.' I asked why. He said, 'Because they're not as smart as I am.' I asked
if that was a requirement for him, because it wasn't a requirement for me.
He said it was, stating, 'With only two of us, I need to make sure I can
learn from the other person in the group. Otherwise, I'm going to be carrying that person.' I asked whether any candidates would be acceptable if
we changed our technical mentoring. 'Oh yes, the first guy was great.' My
developer was trying to make sure that both developers shared the work
and the learning, that one developer was not going to carry the other—a
reasonable concern."

I've worked with other people whose reasons for a veto weren't as sen-
sible and easily addressed. One developer on an interviewing team said,
"I don't want you to hire anyone. I don't want anyone else working with
me. I want to work alone." We found a different project for that person to
work on, and I removed that person from the interview team. If the opposite happens and you have an interviewer who is consistently enthusiastic
about every candidate, ask, "What was most exciting about the candidate
to you?" This question is discussed thoroughly, along with others, in
Chapter 10, "Following Up After the Interview."

Make sure everyone agrees on the work to be done and on the kind of
person who can do it.

On the other hand, you, as the hiring manager, shouldn't make the
decision if you can't obtain hiring consensus or if the situation isn't clear-
cut. If the candidate is ranked "just okay" by the interview team, your best
bet may be to interview more people, rather than hire someone about
whom your group is lukewarm. And don't fall prey to pressure to hire.
Threats such as "If you don't hire someone by March 15, you'll lose the
requisition" are a trap. That's a sucker's game.

Don't train your management to pressure you into hiring someone before you're ready, and don't let an internal candidate's current manager make your hiring decision for you either. Make sure members of your team interview an internal candidate the same way they would interview an external candidate. Sometimes, your colleagues want to unload people and will be less than candid about an internal candidate's appropriateness for your group. Or, they may have the best intentions, but not know about your group's culture and whether the internal candidate will be a good fit for your team.

If you determine that you do not want to consult your interview team—for whatever reason—do not form such a team. That is, if you believe that you should make the hiring decision alone, ask yourself these questions:

1. *Is your competence on the job judged by your ability to make hiring decisions by yourself, without the input of others?* If so, discover who evaluates your job performance and your ability and see whether you can change his or her criteria.
2. *Will anyone else work with this candidate on a daily basis?* If so, why do you not care what your colleague thinks about the candidate?

If you could single-handedly learn everything you needed to know about a candidate, then it would be okay to make a hiring decision yourself. Since that is rarely possible, it makes sense to use the team to help make the hiring decision.

Plan what you will do if you can't find the right people.

Every profession has ups and downs with regard to hiring. During recessions, there may be many candidates from whom to choose. During boom times, the demand for people appears to outstrip the supply. That's when your hiring and management strategies are critical to your success.

You can choose one of several options when you can't find candidates to fill your positions:

• *Expand your search.* Make sure you're taking advantage of all the recruiting possibilities (see Chapter 4, "Sourcing Candidates").

- *Change your hiring strategy.* Hire people who have fewer specific technical skills, but who fit the culture and are fast learners or great problem-solvers—and then train them (see Chapter 15, "Moving Forward").
- *Choose which projects you're not going to do.* Alternatively, choose *when* you will do the projects.

Take a few minutes, and develop a hiring strategy. You'll find the rest of your hiring easier to accomplish. Following is a template to help you develop your hiring strategy:

Problem Categories & Problems to Solve	No	Yes	Desired Characteristics & Problem Solutions
We need more people to do more of the same kinds of projects.			Technical skills, as long as enough candidates exist. If not enough candidates, focus on people's ability to learn and teamwork.
We're making the transition from one kind of technology, work, or product to another.			Problem-solving skills, skills learning new technology, adaptability, and cultural fit.
We're on the cutting edge of technology.			Adaptability, cultural fit, and ability to work in teams.
We're putting together a brand-new group.			Experience working, experience applying functional skills to new product domain, experience creating a new team and making the team successful.
We're filling in with other skills to change what we currently do.			Cultural fit, fit with team, expertise in specific functional skills and ability to apply those skills to new product domain.
We want to make our projects finish faster.			Different functional skills, teamwork, and cultural fit.
We need a few people now, but not forever.			Consider contractors with great communications skills so you won't lose their work when they're gone.
We have to fire too many of the people we hire.			Verify that the interviewing team is composed of people who know how to interview and that they understand the requirements of the position. Use limited consensus to hire people.
Turnover is too high.			Review cultural-fit needs and verify that interview questions address cultural fit.
Recruiting more people is difficult.			Use multiple sourcing mechanisms. Make sure résumé-screening filter isn't too tight.
We need more diversity in our group.			Look for diversity in background, attitude, personality, product experience, as well as in race and gender. Look for different levels of experience.
We need more management capability.			Look for management skills along with cultural fit.

Table 1-2: Hiring Strategy Template.

Once you've listed your concerns, organize them in order of priority to help guide your job definition, recruiting, and hiring actions. Don't forget to explain your objective to anyone who helps you recruit or interview.

Review your checks in the Yes and No columns in the template above. Then choose which actions to take.

POINTS TO REMEMBER

- Know why you're hiring more people. Define your problems to define your hiring strategy.
- Know what types of roles you require. Do you need more developers, more support staff, or more testers? If you had more writers, could you work differently? Are there trade-offs you can make to fill a specific role?
- Know how you will decide on which candidates to select for which jobs. Consider consensus-based hiring as the decision-making mechanism.
- Know that you need a risk-mitigation strategy. If you can't find the people you need when you need them, define what you're going to do.
- Re-evaluate your hiring strategy periodically, based on how much hiring you've completed.

2

Analyzing the Job

Hiring Manager: "I need to hire a developer/tester/writer."

HR Rep: "Okay. What does that person do?"

*Hiring Manager: "Engineering . . . development . . . testing . . . writing,
    of course. A little bit of this, a little bit of that."*

Most technical staff and technical managers do a little bit of this and a little
bit of that, but unfortunately, that's not even close to a job description.

Performing a thorough job analysis is a fundamental component of the
hiring process, and your analysis should help you define the requirements
for the job. As a result of taking the time to analyze the job, you will discover criteria that you can use to create a job description. You won't need
to embarrass yourself by only being able to tell a recruiter you're looking
for "a little bit of this, a little bit of that."

You will want to perform a thorough job analysis if any of the following conditions apply:

- You've never analyzed your open requisitions before in this
  organization.
- You're starting up a group, after a reorganization.
- You need to change the general job descriptions of the people
  you're managing.

If you've defined your *hiring strategy* according to the approach detailed in
Chapter 1, "Developing Your Hiring Strategy," you know the problems
you want to solve. Now it's time to define which kinds of people you need

at what level to solve those problems. You may already have a high-level idea of what you want to see in potential candidates, but performing the job analysis will enable you to see what specific tasks a particular employee will need to be able to perform. In the sections that follow, I discuss job analysis in the context of designers, testers, writers, support staff, and project members in general, but I do not address jobs in management. If you're hiring a technical manager, look at Chapter 14, "Hiring Technical Managers," for how to analyze a manager's job.

Let's start with the high-level tasks you probably already have in your mind. *Developers* may design, implement, peer-review, unit-test, and debug. *Testers* may evaluate designs and perform black-, white-, or gray-box testing. *Writers* may develop new material or they may edit someone else's written material. *Support staff* may answer calls and help customers by determining and fixing problems. It is also possible that the staff you will need may be required to perform different work altogether from the tasks I've listed. To enable you to identify what you want in a candidate, the job analysis must be more than just a description of the job's functions. Your job analysis should identify the reasons you're hiring someone for your particular culture and organization.

Before you start writing a job description, analyze the job in the context of your current team's skills, experience, and personalities, so you know what to look for in a candidate. I use the following steps to analyze the job:

1. *Define the job's requirements,* including defining the person's interactions and the type of work he or she will perform.
2. *Define both the essential and the desirable qualities, preferences, non-technical skills, and technical skills.*
3. *Define the required educational background and the desired level of technical experience.*
4. *Define the activities and deliverables,* the outcomes of the work you want the employee to perform.[1]
5. *Define the factors that could eliminate a candidate* from consideration.

Once I have filled in as much information as I can at each of the five steps, I then complete a job analysis worksheet (a sample is shown in Worksheet 2-1 presented a bit later in this chapter). The worksheet makes it easier for me to iterate on the analysis if necessary later on.

---

[1] For an excellent treatment of performance-based hiring, see Joseph Rosse and Robert Levin, *High-Impact Hiring: A Comprehensive Guide to Performance-Based Hiring* (San Francisco: Jossey-Bass, 1997), pp. 26ff.

Aside from large changes, numerous small changes can sometimes affect your hiring strategy and job analysis. Use smaller changes (such as a change in release cycles or overall volume of work) to trigger a reassessment of your hiring strategy and job analyses. I tend to use the yearly budget process and yearly performance evaluations to review job analyses, especially when I expect to be hiring people during the coming year.[2]

If you've never analyzed a job before, but have existing job descriptions, try filling out Worksheet 2-3 at the end of this chapter, and then match your job descriptions with what you've written in the worksheet. If they don't match, it's time to reanalyze the positions you have open.

One caveat: The perfect person for the job probably does not exist, or, at the least, cannot be hired for what you're willing to pay. That's okay; we all have to live within constraints in a real world. That's why analyzing the job is so critical. The analysis helps you decide which criteria are required and which are optional. You can then find a candidate whose qualities are close enough to the original criteria and choose what tradeoffs you will make in terms of position requirements, qualities, preferences, and skills.

Define the job's requirements.

Begin analyzing the job by creating the high-level view of the tasks to be done and define the kind of person best suited to do those tasks. Think about other people with whom this person will work, and consider the conditions under which they will work together—in effect, the customers or clients of the employee.[3] Identify the personal qualities and preferences of current staff members that enable them to get the work done properly—characteristics such as whether a person is talkative or quiet, or whether someone is innovative or prefers to follow strict guidelines. Define the specific tasks this newly hired person will do, the technical environment in which the person will work, and the kinds of deliverables the person will complete. If you also are looking for specific desirable experience—such as pharmaceutical company experience, recent college training, or the sub-

---

[2] Using your work analysis to define critical attributes can be difficult. If you are an experienced hiring manager, you may have heard of using KSAOs (Knowledge, Skills, Attributes, Other) as a technique for analyzing the job. I don't use them for two reasons: First, I haven't encountered them in any of the technical organizations in which I've helped hiring managers; and second, KSAOs are often underspecified so they are unhelpful to the hiring manager. For a discussion of these, see Rosse and Levin, op. cit., p. 34.

[3] Recognizing that a new hire will interact with users and customers can help you identify the best person to hire. For more on the topic, see Donald C. Gause and Gerald M. Weinberg, *Exploring Requirements: Quality Before Design* (New York: Dorset House Publishing, 1989), pp. 68ff.

contracting experience a military contractor might have—list that on the job analysis sheet. Or, if this person may need specific talents and skills to complete the work in your time frame, list that detail as well. When you spend time analyzing the job at the start of the hiring process, your recruiting and screening efforts are more effective.

List the job requirements to include five things: (1) interactions, (2) functional roles, (3) role level, (4) management component, and (5) activities and deliverables.

I find that if I start by defining with whom the person will work, before I list the candidate's deliverables, I'm more likely to develop a job description that I don't need to revise after I've seen the first few candidates, wasting precious phone-screen and interview time. No matter how you start an analysis, don't forget to include all the analysis components in your position definition.

This approach admittedly is a top–down approach, and may not be comfortable for you if you prefer to start with the details. If you prefer, try working from the inside out, starting with the role level, the role, and the interactions. Then, define the management component, activities, and deliverables. Don't start with the activities and deliverables first; you'll forget something crucial in the interactions, role, and level.

To define the requirements for candidates, use the following techniques:

First, define the expected interactions. Define with whom this person works. One way to think about the interactions is to consider who the person's customers are, and who the suppliers are. To whom does this person deliver work product or information? Define the daily, weekly, and monthly team and personal interactions. I tend to use titles of people, not names here, so that my job analysis is clear even if the group experiences turnover. However, if you do use actual names, the people named may be willing to help you review the analysis. Write down how the person in this position will work with other people: cooperating, influencing, providing work direction, negotiating, or working as part of a team. Specify the frequency of the interactions.

Second, define the functional roles for the position. Sometimes, you only need to record "developer" or "tester" here. But if you have a writer who also fields calls, the position is a combination of roles. I like to think of functional activities (architecture, design, programming, planning, analysis, quality assurance, release engineering, integration, documentation, and testing, for example) when I define the roles this position requires. Then, go on to define how large a part each role plays in the position.

If you're defining a job with multiple roles, spend enough time on this step to know which roles are most important, or how much time you want each role to take. People find it difficult to work when they have to spend time on two separate functions. If you're looking for a support representative who will also test, a writer who will also support, or a manager who will also perform development, decide what percentage of time you think is appropriate for each function, based on the activities and deliverables you need from the position. Keep in mind that once you've hired someone, the percentage may change.

Third, define the position's level. You may find it difficult to determine whether this is a senior- or junior-level position, especially if your organization does not have a formal, posted job ladder. I follow the guideline that the more experience and the broader the experience required, the more senior the position. When in doubt, look at the deliverables and activities, discussed below.

You might want the new person to change the balance of seniority in the group. If your group is top-heavy with senior people, you may want to start adding junior people who can be trained. You can make good use of a junior-level person even if the position is on a time-critical project that requires a lot of technical knowledge, simply by moving a current senior-level staff member to the new position and backfilling his or her current work with the new junior-level hire.

Fourth, define the position's management component. Many technical people take on technical leadership roles at some point in each project. However, some positions are combinations of technical work and technical leadership. Or, the position may have elements of project management or people management. Decide now if this is a purely technical job.

When you define a management job, define the scope of the job. For example, is the scope strategic (requiring a person with the ability to plan an organization's operations), operational (requiring someone with the ability to plan and oversee the daily work of the organization), or supervisory (requiring a person who has demonstrated administrative ability and who is task-oriented)? Then, define which functional area or areas are involved in the job. Analyzing a manager's job is more difficult than analyzing a purely technical job. See Chapter 14, "Hiring Technical Managers," for more information.

Fifth, define the position's activities and deliverables. Most people start with this fifth task when writing a job description, but I caution against jumping in without having done the previous tasks. Although it is true that a developer develops, a tester tests, and a writer writes, and that for some jobs, the activities and deliverables are that easy to define, most technical positions require more than just a one-word functional description of activities and deliverables.[4] The more specific you can make the deliverables, the better your job analysis will be. If you are hiring for specific work, be specific in the activities and deliverables, noting for example, "Complete the Big Project by Jan. 30," or "Lead the architecture effort for ModuleX, completing the initial architecture by June 5." The more specific your activities and deliverables are, the more people on your interview team can focus their questions.[5]

Developers may perform requirements elicitation, requirements analysis, model building, design, architecture, analysis of defects found, coding, debugging, unit-testing, and more—for a specific project.

Testers may perform test planning, test development, and defect report generation; they may gather and disseminate metrics reports about defects or performance, and more—for a specific project.

Writers may write documentation, edit, run tests, develop examples, develop tutorials, and more—for a specific project.

In addition to project work, each position has daily, weekly, monthly, and some yearly deliverables. When you consider deliverables, consider how you will measure the employee's success. If you think about what success means in this position, you will be able to define the activities and deliverables more easily. Be as specific as you can, using completion dates, module names, product or project names, or people's names. For example, by specifying "Mentor junior members of the tech. pub. group to design a new document for the BuyWrite project by December," you have defined both activities and deliverables for the open writer position. Write the activities and deliverables as if you were writing a yearly list of goals and objectives for an already hired employee.

Some position's deliverables are harder to define. If you need a technical coach or a facilitator, describe those services in terms of the benefit to the position's customer. One VP of engineering enjoyed benefits he attributed to his agile-programming coach's services: "He watched how people worked together. He looked for teams that weren't clicking. He made them click in a way that made the project manager happy." The VP also could have described the agile-programming coach's activities and deliverables this way: "Monitor the team—daily. Look for problems the team has

---

[4] See Rosse and Levin, op. cit., p. 27.

[5] An excellent treatment of the importance of identifying the critical performance objectives of every job appears in Lou Adler, *Hire with Your Head: Using Power Hiring to Build Great Companies,* 2nd ed. (New York: John Wiley & Sons, 2002), pp. 26ff.

using the methodology and other obstacles. Remove any obstacles to the team's success—daily." Here, the position is described in terms of benefits to the team members and project manager.

I find that as I define activities and deliverables, I look repeatedly at and review the other parts of the analysis, specifically the roles and the level. Defining the level of seniority needed for whoever fills the position may help you judge the activities and deliverables more easily.

Job Category	Activities and Deliverables Needed to Satisfy Job Requirements	
	Junior Level	Senior Level
Systems Engineer	Analyzes current requirements, looking for ambiguity.	Develops requirements cooperatively with others; assesses requirements documents for their completeness, fit with the rest of the system, and for ambiguity.
Developer	Designs modules after the major portion of the design is complete. Writes and compiles code in accord with some predetermined description. Arranges to have his or her own designs or code reviewed. Reviews peers' code. Uses configuration management system to check code.	Moderates code inspections, drafts design specifications, drafts functional specifications, and designs large pieces of the system.
Tester	Develops test procedures. Runs tests and reports on tests. Gathers metrics. Automates pieces of test procedures. Attends reviews and inspections.	Designs test approach. Designs automated test procedures. Moderates reviews and inspections. Develops metrics reports.
Writer	Writes documentation.	Plans tech. pub. and book design. Plans on-line help design.
Support Staff	Takes information for each incident. Resolves incidents that don't require looking at the code.	Resolves any incident. Manages customer expectations for fixes. Manages the fix process for urgent problems.
Release Engineer	Writes scripts. Creates builds.	Sets up the configuration management tool. Manages and anticipates storage needs.
Project Manager	Plans and organizes small projects. (Make sure you define what small means to you: four people and four months, fifteen people and eighteen months, or something else?) Performs basic risk identification.	Manages large-scale programs and systems, bringing project managers across the organization together to deliver an entire product, not just the technical part of the product. Negotiates with suppliers and customers.
Manager	Facilitates problem-solving. Performs strategic planning. Performs risk management.	Manages multiple groups. Coaches and mentors other managers.

Table 2-1: Junior- and Senior-Level Activities and Deliverables.

I make the distinction regarding junior- or senior-level requirements when I determine how much the position is worth to the company. The more senior the position, the more it's worth. For example, I'd expect to get a less-professional functional specification from a recent college graduate than I would expect from a senior architect. When I choose the level, I must be willing to pay for the work completed at that level.

If the company isn't able to pay what you think the position is worth, then be clear on the deliverables you can expect from possible candidates. If you expect too much of your candidates without offering enough pay, you may never hire anyone. Candidates will go through with the interview, conclude that you are asking too much for too little pay, and decline your offer.

Once you've described the deliverables, you're ready to look at the other necessary candidate characteristics.

Define the essential and desirable qualities, preferences, and non-technical skills for a successful fit.

The second part of your job analysis is to look at the qualities, preferences, and skills that will help a candidate succeed in the position and fit into your organization's culture.

Different companies can have different cultures. Take, for example, two organizations that are making high-performance commercial products. The first organization, SpeedyOne, is a typical entrepreneurial organization in which self-starting, responsibility-taking, and the ability to do three things at once are highly valued. Risk-takers are especially successful at SpeedyOne.

SpeedyTwo is an older organization whose emergence from start-up to established company was rocky, a factor that led to a significant aversion to risk in senior management. SpeedyTwo values employees with a passion for learning, the ability to make decisions by consensus, and the focus to finish projects. To an outsider, it would appear that both companies require similar technical skills. However, people at SpeedyOne prefer to work by taking initiative and forcing products to market quickly. The technical staff at SpeedyTwo uses consensus to make decisions, and uses technical practices such as peer reviews and walkthroughs to move their products to market. Each company looks successful and meets time-to-market needs, but the cultures differ dramatically.

There's nothing predominately right or wrong with how business is conducted at either company. The point to be made is that hiring managers must differentiate between the types of people who'll be successful in vastly different cultures.

Sometimes, the main reason a hiring manager doesn't hire a candidate is that he or she has a gut feeling that the person just won't fit well with the culture. But a "gut feeling" is not a good reason not to hire someone, so train yourself to articulate culture-fit differences. If you can't articulate why a person won't fit, you run the significant chance of hiring someone else with similar problems. Or worse, you hire people and then are vaguely dissatisfied with their performance, having expected "more at this level." If you have trouble articulating why candidates are not quite appropriate, then defining the qualities, preferences and skills *prior to hiring* will help. If you're dissatisfied with some of your recent hires, take the time now to define the essential qualities, preferences, and skills that will fit your culture.[6]

Think also about diversity with respect to culture-fit issues. Sometimes, your group can be too homogeneous, so you want people who *don't* currently fit.

Essential qualities, preferences, and skills

A well-chosen employee who is successful in performing the job for which he or she was hired almost always will fit into the team and the larger organization. In addition, he or she will meet the technical challenge as well as fit the culture.

No matter what size a company is, culture can vary dramatically across the organization. You need to define the cultural needs as well as the technical needs for your immediate area, regardless of whether it sits within a corporation that is large or one that is small. Sometimes, a candidate who would be a great fit for one manager or in one organization would be a disaster for another simply because different managers and different organizations have their own unique styles. Review your group's culture to see which qualities are present in successful employees in your group, and compare how well candidates for the position match up.

Start your search for technical and cultural compatibility by assessing each candidate's *qualities,* such as initiative, flexibility, and technical leadership*:*

- **Initiative:** Do you need someone who looks for problems and fixes them? Do you want someone who is intellectually

---

[6] A thorough guide to defining culture-fit compatibility is Jim Harris and Joan Brannick, *Finding & Keeping Great Employees* (New York: AMA Publications, 1999).

curious? Do you need someone who can follow directions. Do you need someone you can train to do more and different work? *Define the amount and type of initiative you and your organization require from an employee.*

- **Flexibility:** Do you want someone who is completely flexible? Do you want someone who will develop rules of operation for you? Do you want more flexibility in a senior-level help-desk representative than you would require in a junior-level software developer? *Define the flexibility or adaptability required in the position.*

- **Technical leadership:** How much leadership do you want the new hire to take on? Do you want someone who is capable of creating new ideas, someone capable of sifting through ideas to discover the most appropriate solution, someone who will catalyze people to generate new ideas or decide on an idea, or someone who can follow through with the details so that an idea comes to fruition in a product? *Define the kinds of technical leadership you need from the candidate, keeping in mind that technical leadership does not necessarily correlate with years of experience and that not every employee needs to be a leader.*

After you have considered necessary qualities, consider a person's *preferences* for how he or she likes to work, and match them to your own preferences for how the job is to be done:

- **Procedural preferences:** Is it important that specific procedures be always followed to the letter on the job? Do you need people who take exceptional pride in following procedures? Can you tolerate mavericks who live to break the rules? *Define what procedural tolerance level you need.*

- **Tasking preferences:** Do you need a multitasker who likes to work on multiple tasks at one time, juggling, say, six tasks in various stages of "done-ness," or do you want someone who likes to handle only one at a time? Do you want someone who works to complete tasks, or a person who is happiest when required to context-switch between multiple projects or tasks? Do you want someone who can handle uncertainty, or a person who needs well-defined limits and schedules? *Define your tasking needs, but keep in mind that* *when looking for a multitasker, you want someone who can let you know when he or she can't take on additional work.*

- **Goal-oriented preferences:** Do you want someone who can set and reach his or her own goals without much input from you or others? Can you deal with people who want to set their own goals rather than look to you to set the goals? If you prefer not to be a hands-on manager and choose to let goals evolve, can you manage someone who needs specific and detailed goals? *Define the goal orientation you require in an employee.*

- **Problem-solving preferences:** To what degree do you want people to own and solve their problems before they bring them to you? Do you want someone who will ask you to help establish task priority when confronted with conflicting tasks or schedules? *Decide how much independence is appropriate for the job and for your group.*

- **Learning preferences:** How much initiative do you want someone to take to stay current in his or her field? Does staying current matter for the particular job? *Determine whether the employee needs to have a yearning to stay up-to-date in all areas of technology or whether current skills are sufficient.*

- **Collaboration preferences:** Do you need a person who prefers to work alone or one who thrives when working with a group of people? Do you need a catalyst for a team? Are you looking for someone to complement the team? Does the position involve a significant amount of group work or very little work with teammates? *Define the amount of collaboration you require in the position.*

Finally, consider the *non-technical skills* that might make a person successful in your group:

- **Communications skills:** Do you need someone with excellent speaking or writing skills, or both? Do you want someone you can put in front of customers, a person who will be quick-thinking, good at fielding off-the-cuff questions, and the like? Do you require excellent phone skills for the job? *Determine your non-technical needs in the context of the specific job the person is being interviewed for—be cautioned, of course, about the legal ramifications of seeming to discriminate* *against someone who lacks specific non-technical skills when such skills are not required for the job.*

- **Performance-versatility skills:** Do you need someone who is strictly tactical and operational, or someone who can think strategically and plan what has to happen? Do you need a person who can handle projects of varying scope or a person who wants to focus on one kind of work? *Determine whether this person needs a variety of problem-solving skills, or a narrow range of skills focused in one product area or one kind of project.*

- **Negotiation skills:** Does the person need to be able to work with people inside and outside the group (or the company)? Is there management by authority, by influence, or both? Does this person need to manage choices between competing designs? Do you need someone who can negotiate with potential customers or different groups within the company when defining requirements? *Determine what level of skill your new hire will need as a negotiator.*

- **Problem-solving skills:** Will the new hire need to think about problems in a variety of ways? Can you use someone who takes the first solution that presents itself? How much creativity do you need in this role? *Determine what problem-solving skills are needed for the position.*

Not all the cultural qualities, preferences, and skills mentioned here will have the same degree of importance to you as they do to someone else. You may decide to select other qualities, preferences, and skills that your technical staff needs in order to succeed. Think about the successful people currently working in your organization and identify the qualities, preferences, and skills you find most valuable.

Each corporate culture is different, so define your essential qualities, preferences, and non-technical skills for your open position. You may realize that you have a particular type of person you choose to work with.

Following is a sample worksheet that can be used as you define the qualities, preferences, and skills required in a particular job. Add other characteristics required or desired for the job, using the Notes column to describe how those qualities, preferences, and skills fit the requirements of the job.[6]

---

[6] For talents that you may want to include when you define qualities, preferences, and non-technical skills, see Marcus Buckingham and Curt Coffman, *First, Break All The Rules: What the World's Greatest Managers Do Differently* (New York: Simon & Schuster, 1999), pp. 251ff.

Quality, Preference, or Skill	Required	Desirable	Notes (Cite any required quality, preference, or skill specific to the job.)
Quality: Initiative			
Quality: Flexibility			*This project manager needs to manage projects that require different lifecycles.*
Quality: Technical leadership			
Quality: Responsibility and independence			
Preference: Ability to work on multiple projects at one time			*The tester needs to be able to juggle planning for one project while developing tests for another project.*
Preference: Goal orientation			
Preference: Passion for learning			*We need someone who wants to keep up with the literature as new things are happening in the field all the time.*
Preference: Teamwork			
Skill: Communications skills			
Skill: Ability to handle projects of varying scope			
Skill: Influence and negotiation skills			
Skill: Problem-solving skills			
Add your qualities, preferences, and skills here.			**Add your notes here.**

Worksheet 2-1: Matching Qualities, Preferences, and Skills with Job Openings.

Desirable qualities, preferences, and skills

For many open positions, some of these qualities, preferences, and skills are not essential, just desirable. As you analyze the job, decide whether any of these factors would make you reconsider a candidate's pay, level of experience, or cultural fit in terms of the position for which you're hiring.

For example, technical leadership might be an essential quality in candidates you're recruiting for a position. The ability to help you develop and present project or program status to management would be desirable, and you'd be willing to pay at the high end of your salary range to get it.

Identify corporate cultural-fit factors.

Once you've fully described the qualities, preferences, and skills you want in a candidate, think about cultural fit. Your company has characteristics specific to it that will make or break a candidate's cultural fit. Not every organization is perfect for every person—and vice versa.

Think about your company's factors as you continue preparing your job analysis worksheet. People choose to work for companies for a variety of reasons, some of which have little to do with the technical work of the specific job.

Possible company-fit factors appear in Table 2-2, below.

Company Fit-Factors	Preference Possibilities
Working environment	Some candidates like offices with doors so they can have discussions with others without disturbing their colleagues. Some candidates prefer cubicles or a bullpen office, where people are accessible.
Career growth paths	Candidates may care about upward mobility in your company after they learn about your products and produce good results for you. If your company has not yet developed a technical or managerial job ladder, suggest that your Human Resources or Personnel staff do so now. Be ready to explain to a candidate your company's growth paths. Some people like to work where they can acquire more skills, prove themselves valuable, and then change roles to the next phase of their career at the same company. If this is beyond your scope or influence, ask your boss what to say to candidates.
Start-up or established	Some candidates only work for start-ups. Some only work for established companies.
Your products	Some people won't work for a company that produces tobacco products or munitions or parts that end up in weapons. Some people don't care about the product as long as it has a GUI or a scheduler or a compiler, for example. Some people only care about working in a specific industry, or on cutting-edge technical products, on consumer products, on end-user products, or on technical-user products.
Industry leadership	Some candidates like to work for industry leaders, or companies they think will be industry leaders soon.
Training policy	Some candidates look for reimbursement for education or other training opportunities.
Competing local employers	If you are one of only a few employers in a small town, it may be that none of these factors weigh as heavily for a candidate as the possibility of steady employment.
Company's profitability	Some people like to work for companies that are floundering because they enjoy the challenge of turning a company around. Some people prefer to work for a company that has a healthy cash flow and provides more security.
Company growth	Many candidates look for positive growth because there is the assumption their jobs will grow and they will continue to be paid. If the company's growth is stalled or negative, you can attract candidates who pride themselves on their ability to come into a negative situation and make it positive. Some candidates will be oblivious.
Cash flow	If the company is wealthy, candidates may assume that the company will buy necessary tools and provide training.
Recent layoffs	Some candidates will not interview at a company that's recently laid off large numbers of employees. Some candidates like the idea of a fresh start.
Recent acquisition or merger	Some candidates see acquisitions and mergers as a time of reduced productivity and increased drudgery. Others see them as a time of opportunity.
Overall corporate culture	Some cultures are dynamic and appear chaotic. Other cultures are calmer and more relaxed. Some candidates choose to work for a company because the projects and the organization are everchanging. Others choose to work for a competing company because the products and organizational structure is stable.

Table 2-2: Company-Fit Factors.

It's common among technical testers and support staff to want to move into development. I've hired people into testing or support, trained them for six months, and then let them move on. Other hiring managers want a guarantee that the employee will be in their group for twelve or even eighteen months. As a seasoned hiring manager, I'm willing to live with having an employee in my group for fewer months because I've had the chance to train the employee and I can use him or her to help me interview for a replacement. Whatever your position is, make sure you can articulate it, and that your position matches your company's policy on job changes.

One company's employment policy may be attractive to some candidates and repulsive to others. I once worked for a company that would not permit smoking anywhere on its premises. I found this attractive, but the smokers I phone-screened did not appreciate the policy.

Some companies expect their employees to participate in a variety of extracurricular activities, such as raising money for charitable organizations, weekend sales meetings, social events, and so on. Candidates who find these activities exciting, and view them as a positive use of their time, are an appropriate fit for those companies. Candidates who do not want to participate in these extra-corporate activities are not a good fit. Avoid wasting your time interviewing people who aren't a good cultural fit, especially if the lack of fit is a misalignment of cultural values regarding where employees should spend their time.

If a candidate's success hinges on these factors, then remember to ask questions about these specific cultural factors during the phone-screen and also during the interview. See Chapter 7, "Developing Interview Questions and Techniques," for suggestions.

Define the corporate cultural-fit factors so that you can use them in a job advertisement. You'll have more success attracting the most suitable candidates if you identify such factors right at the start. Even if you don't want to use the factors in an ad, being aware of the cultural-fit factors will help you answer candidates' questions about the company.

Define the necessary technical-skill level and the required educational background.

The third part of the analysis involves defining the required minimum educational background a candidate needs to be successful in your organization, and the minimum technical skills required for the candidate to complete the deliverables. Consider the work you need performed and the background required to perform it. Determine if your organization has assumptions about required educational background for technical people.

Prior work experience

A candidate with *prior work experience* can be very valuable, but you'll want to look beyond what is stated on the résumé. Define the work experience required of a candidate before you consider him or her for the job. Consider functional skills, product domain, tools and technology, and industry experience.

Functional skills experience

A candidate who has *functional skills experience* possesses a technical understanding of how to perform the work required in the position. When you define functional skills experience, define what evidence you'll require to confirm that the skills you're looking for are present.

For developers, functional skills include proficiency in reading and writing code, as well as skills in designing, programming, and debugging. For testers, functional skills include boundary-condition testing and equivalence partitioning. For writers, functional skills include proficiency in grammar. For project managers, functional skills include knowing how to apply lifecycles to a project, as well as how to estimate, measure key project metrics, and run a meeting.

Functional skills experience is both technology-independent and tools-independent. Functional skills experience is gained in school and on the job. However, candidates may need specific functional skills to be successful in *your* environment, such as understanding how database schema are developed and used. Or, you may require a person with varied knowledge of data structures, or how to create and use watchdog timers, or with an understanding of how to avoid race conditions in real-time systems. If the position requires specific functional skills, include those skills in the activities and deliverables, for example, noting that you need someone to "design and implement the watchdog timer for ModuleA by June 1."

Product-domain experience

For a candidate to have *product-domain experience*, he or she must have knowledge of the product—either of the problems the customers want solved by this product, or the internal architecture of the product, or both. You may want someone who already understands how the product works and why your customers use the product. Or, you may want someone new to this kind of product, so that you receive the benefit of fresh eyes and attitude. For example, when I worked in machine-vision companies (companies whose products use cameras and software to "see"), our goal was to hire developers with machine-vision experience, so that they already understood the domain issues. When we wanted to hire testers, we needed people who had real-time and embedded systems experience, but not necessarily machine-vision experience, because we wanted the testers to be open to alternative testing possibilities.

When you assess your hiring needs, judge each candidate's product-domain experience in the context of your products. When you define a position and assess candidates, judge their experience against the products they've already worked on.

I identify product-domain experience as either problem-space or solution-space:

- With *problem-space domain expertise,* the candidate understands how the users use the product. Candidates generally have exposure to and a general understanding of product externals. They understand how the product interacts with the users or, from a black-box level, how the products' interfaces work to solve the customer's problem. Technical people acquire this level of expertise quickly.
- With *solution-space domain expertise,* the candidate understands how the product works on the inside—that is, the architecture of the product. We expect developers to have this expertise; if they don't, we expect they can acquire it by reading the code. Writers can acquire this expertise, depending on the kind of documentation they write. Testers and technical support staff members can acquire this kind of expertise if they can read how the product works on the inside, or if they are taught about the internal workings.

A candidate who has the ability to acquire both problem-space and solution-space expertise is likely to be most valuable to you. As you define the job (and later, as you review résumés or interview candidates), determine how much expertise the candidate must have both in how the product works (solution-space expertise) and in how to solve the problems of the customer (problem-space) using his or her functional skills. People with both kinds of domain experience understand the specifics of how the product works, and can understand how to relate their functional skills to improve the product. At the senior level, developers, testers, designers, and architects all are likely to have this expertise. It's easier to obtain this level of expertise if the candidate already can read and write code, but it's not required. All technical staff members can acquire both problem-space and solution-space domain experience if they have the ability to understand the product and are educated about it.

How the different kinds of domain expertise work together is illustrated in the following example: If you have a multi-threaded product, one in which numerous instances of the product can be run simultaneously on one machine for multiple users, then a candidate has reached problem-space expertise if he or she knows that the product is multi-threaded and why the product requires multi-threading (for performance or throughput or whatever). With solution-space expertise, the candidate knows what kicks off multiple threads, and understands how to use that knowledge to develop pieces of the product appropriately or to create tests that test the product differently than if the product were single-threaded.

Technology and tools experience

Having experience with the hiring company's specific programming language, operating system, database, or other tools gives a candidate somewhat of an advantage over applicants without suitable *technology and tools experience*. Mastery of the specific technology and tools can be easily taught to a candidate who is fully qualified for the job in all other areas. However, technology and tools experience may or may not be a consideration in whether you or your company will decide to hire a candidate. The decision more likely will depend on how quickly you want him or her to start being productive in your environment. For example, if you're starting up a company that will use an Oracle database for a server and a Java front-end for a client, you want to hire people who know how to use Oracle databases and Java front-ends, as well as how to use the specific architecture to design, develop, test, document, and support that kind of a product.

If you're not doing the phone-screens and interviewing yourself, make sure you specify to the people who are performing those tasks whether you need to hire a person with experience using a specific tool or someone with general tool experience. For example, if you're looking for a person to perform test automation, experience with any of the test automation tools may be acceptable. However, if you're looking for a person to coach your team members as they write automation using a specific tool, then experience in that tool is probably necessary. Similarly, a project manager with experience using any of the project scheduling tools may be acceptable, unless you also want that project manager to teach new project managers how to use the tool. Teaching, coaching, and mentoring activities may require a good working knowledge of a specific tool. Otherwise, general tool experience is probably sufficient.

Industry experience

A candidate who has *industry experience* understands not just who your customers are, but how your customers will react, and what they expect from their systems. Industry experience relates to how people use the product; product-domain expertise is knowledge of the products' internals.

You may want to hire a candidate with experience specific to your industry so that he or she understands your customers and their expectations, and has an understanding of the types of problems you encounter in your work. For example, people who work on software for airplanes understand that their products will be audited and their processes will be assessed to ensure that the product development group hasn't created an unsafe environment. Someone who's worked in the shrink-wrap, commercial-product-development world might not welcome all these assessments and audits, but a veteran of the aeronautical industry must be more accepting of them. The pharmaceutical industry is another example of an industry in which process rigor and audits are common practice.

Focus on the experience necessary for the job you have to fill. Don't worry about planning too far ahead—it's too hard to predict the future. I've found that when I plan too far into the future, my candidates don't have the skills to perform the work that I need done now.

Identify essential technical skills.

Only you know whether it is important to fill the current open position with a candidate who will know how to use your technology immediately. We all spend time training people to be successful in our organizations, but are you planning specific skill-based training in addition to helping the person navigate the unfamiliar seas of your workplace?

When you define the technical skills required, make sure you know what's actually required for the particular job. For example, if the product is written in Visual C++, you may require someone with a number of years of Visual C++ experience. The number of years of experience you require should depend on whether this is a senior- or junior-level role. If a working knowledge rather than in-depth experience is required, then you may not need to specify a minimum number of years of experience in the job analysis worksheet and the job description. However, if you are seeking a mature candidate who has worked on numerous products, then specify overall years of experience. If you're seeking a person with in-depth technical knowledge, then look for someone with a few years of specific experience instead of a variety of technical skills spread over the years. One way to define the required technical knowledge is to ask your current team members to help you define the requirements.

When specifying technology, remember to consider your specific development environments. Someone who has developed software in the C language using a UNIX operating system may have a different idea of how to develop software than someone who has developed software using Visual C++. In the job analysis, the job description, and also in any advertisements, specify the minimum number of years of experience you want an applicant to have in each environment.

When you consider functional skills or product-domain experience, think about whether you want someone with experience throughout the entire project lifecycle. Someone who's lived through a product release will have had a different experience from someone who's worked only on canceled projects or who has been moved off projects before their completion.

Carefully consider which skills you need a candidate to possess, and which skills you are willing to provide by training the person after he or she has become an employee. In a highly competitive job market, it may make sense to hire candidates who have appropriate problem-solving skills, who are adaptable, and who demonstrate an ability to learn, but who don't necessarily have experience in the specific operating system or programming language they'll need to use. In most cases, you will be able to train them in the required programming language in less time than it could take to wait for just the right candidate to cross the threshold. In a less-competitive job market, if you have specifically described the necessary technical skills in the job analysis, job description, and all advertisements, you will reduce your recruiting time because you will narrow your field to appropriate candidates only.

Avoid the appearance of requiring applicants who have more experience with a specific tool or technology than you can reasonably expect or than you truly need. When object-oriented programming came into vogue, reliable, commercial compilers had only been on the market for about a year, but some companies were requiring job applicants to have a minimum of five years of C++ experience. This kind of unreasonable requirement only encourages candidates and external recruiters to stretch the truth, or equally problematic, to not send you their résumés. You, and anyone else involved in describing an open position, need to learn enough about how your company uses technology to hire people for your group.

It's easy to use years of experience as a shorthand indicator of the experience or knowledge a candidate would need to be successful in a position. Nevertheless, when you analyze the position's needs in terms of work experience and technical skills, you must ask yourself what you mean by "work experience."

When you've defined your technical experience requirements, add them to the bottom-most boxes in the job analysis form, as shown in the worksheet portion replicated from Worksheet 2-1:

Quality, Preference, or Skill	Required	Desirable	Notes *(Cite any required quality, preference, or skill specific to the job.)*
Skills: Technical. One year of ClearCase administrative experience.	Required		*We use ClearCase and do not have the budget to train a novice admin. for this position.*

Worksheet 2-1 (continued):
Matching Qualities, Preferences, and Skills with Job Openings.

Technical skills confusion?

You may sometimes need to fill a position that is common at other companies, but that does not already exist at your company. Or, perhaps you know some of what you want done in a job, but not enough of the requirements to feel comfortable about listing the job opening. Perhaps you need to hire a technical support manager, and you know something about technical support, but you do not know enough to describe the essential skills required. If you are faced with such a problem, try the following approach:

- *Ask for help from someone who might know what the job entails.* Such a person might be someone who either has done the work or has successfully hired people for the job—other managers at your company, or outside colleagues, for example. If you're using internal or external recruiters, ask them for help. Ask for advice from consultants, an academic advisor, your mentor, or from someone whose Web page interests you and is relevant to your industry or technology. The more people you ask for information, the better able you'll be to refine the job analysis and the closer you will get to determining the essential job functions.
- *Use analogy.* If you don't know what job a particular job title describes, approach the job analysis from the point of view of the tasks that need to be done. For example, software companies began using configuration management systems in the 1980s. In the early 1990s, the technology was still new

to the software community, and there were not enough release engineers and configuration managers to fill the open jobs. However, people did know what tasks needed to be done to create bills-of-material for software products, for example. So, while these managers were not called release engineers or build engineers or configuration management engineers, they knew what they needed to perform and were able to define the essential job functions.

- *Ask your Human Resources or Personnel Department staff members to compare the offered salary and compensation package with that printed in industry-wide surveys.* Companies that participate in salary surveys have access to lists of job titles and job functions. You may be able to use these lists as a place to start defining the essential technical skills.

Identify desirable technical skills.

While analyzing the job, you've probably thought, "I'd like this-and-that technical skill, but I don't require it." Now you can add those desirable skills to the job analysis.

A True Story

Donna, a Tech. Publications manager, was defining a writing position. She had already determined that a desirable skill was "project-management experience." The writer was going to work on a project in which the project manager was stretched too thin, and none of the other groups had people who could help with management of the project. Donna was ready to pay more in salary if she found a writer who could also help the project manager manage the project. Donna listed "project scheduling and coordination" as desirable skills.

The moral: When defining desirable skills, don't forget to consider your tradeoffs. You may want to change the job level and salary depending on whether you find a candidate with more or fewer desirable skills.

Evaluate educational or training requirements.

Doesn't every technical person need a college degree, at the least? No. A college degree shows a kind of perseverance, but not necessarily the kind of perseverance you need in the person you hire to perform your work. College degrees awarded in the fields of science, engineering, or computer science indicate that the candidate may have learned enough technical information to understand the job to be performed, but degrees don't mean the candidate can perform as needed. Don't let the presence of a degree convince you that a person has the skills and characteristics to do the job well. And don't let the lack of a degree deter you from screening and interviewing people whose experience looks like it might fit your opening.

A candidate's experience with successful product development, release, and support can be more valuable to you than a college degree. One of the best test developers and all-around system administrators I ever worked with was someone who began programming at the age of eighteen and didn't bother going to college until he'd reached his late twenties. Some of the best developers I've worked with never graduated from college. Many good managers I know never even finished college, let alone obtained an MBA or any other advanced degree.

If your HR Department has a policy against technical candidates without degrees, talk to anyone who will listen to discover whether degrees are shorthand for describing some level of competence or experience. (In my view, the real key is whether the person has learned anything from whatever education or experience he or she has had.) Then, decide whether to either fight the policy or live with the restriction.

Experience catches up with formal education. I have observed that technical-degree holders lose technical proficiency if they don't use the particular skill or don't keep up with advances and changes in the field. Remember, too, that not all schools teach the latest technologies, practices, and techniques in their undergraduate curricula.

Depending on your culture, academic credentials may be essential. If you're hiring for a research environment, and your internal customers measure your staff members by the degrees they hold, look for people with degrees. If you're creating a professional services organization, and external customers will want to know how many and what kinds of degrees staff members have, your staff members may need a surfeit of advanced degrees.

A candidate's education or training only tells you the candidate has gone to school. What you want is to find the candidate who has learned to think. Education or training doesn't tell you whether the candidate has learned anything applicable to your job.

Licenses and certifications

What should you know about licenses and certifications? The first fact to remember is that governing bodies—usually the state or city in which the skill is practiced—oversee professional licenses, obliging the licensee to assume legal responsibility for work performed under the license. For example, a licensed "Engineer" has a legal responsibility for the quality of any design he or she signs off on. If you require someone with a license, then specify that license in the job analysis.

Professional organizations, rather than governments, oversee certification, but certification confers no legal guarantee for the quality of products produced by certificate holders. Certification is an indication that the candidate has experience in the field, was motivated enough to pursue the certification, and has mastered enough material to pass an exam. Unfortunately, even though certification may require work experience, the bodies that grant certification don't verify that the work done by the certification holder was successful or is even applicable to your needs. Certification does indicate that the candidate has learned the material well enough to pass an exam, but it carries no guarantee that the candidate can apply the knowledge to his or her work.

I personally do not consider certification to mean anything much when I am hiring someone for a technical position. Because the knowledge tested is functional-skills book knowledge, make sure you understand what the person must do to maintain his or her certification and the value of that certification to your environment. Also, determine if you will need to make any accommodations for the employee to maintain the certification. Many certifications require ongoing education in some form, so you need to be clear who will pay for that.

List the license or certification requirement as part of your job analysis if you do require either or both.

Define all elimination factors.

Now that you've described the job requirements, consider other requirements that would *eliminate* candidates from successfully performing the open job.

Each position may have factors that eliminate otherwise suitable candidates, that is, candidates who otherwise have the necessary technical skills and fit the culture. If you include these factors in your job analysis, you can build them into your job description, and avoid interviewing someone who seems perfect but can't meet some of the non-technical requirements of the job. Frequently, elimination factors fall into the job and company categories shown in Worksheet 2-2, below:

Possible Elimination Factor	Elimination Category	Issue
Travel	Job	Do you need someone who can travel half the time or even part of the time?
Availability	Job	Do you need someone who can commit to core hours or off-hours? Do you need someone who can rotate time on the telephone (common to technical support jobs), such as evenings or weekends, not just during the day? Do you need someone who is at work by a fixed time each day?
Relocation	Company	Are you willing to pay for the candidate's relocation? If you've posted the requisition on the Internet, people outside your geographical area may send you résumés. Some people will pay for their own relocation, so don't automatically reject candidates outside your immediate geographic location, but decide in advance whether you will pay for relocation.
Personal Characteristics	Company	Are there people your company chooses to legally discriminate against? As long as who you don't want to hire is not part of a protected class, such as people older than forty or a member of a minority group, then you can make this characteristic an elimination factor. If your company has a specific reality to consider, then you can use this reality to eliminate potential candidates who do not fit.
	Job	For technical support people who will handle telephone help-desk calls, the ability to speak clearly and audibly is frequently a plus. Decide whether you are willing to train people to speak more clearly, or whether you want to avoid interviewing them altogether.
Salary Requirements	Job	Salary requirements, additional benefits, and other perquisites should be defined in the job analysis and discussed with the candidate during the phone-screen so that such components of the job offer do not become a problem.

Worksheet 2-2: Job and Company Candidate-Elimination Categories.

Define the elimination factors for your job on the job analysis worksheet, and use the elimination factors in your phone-screen, sometimes as the first questions you ask.

Think twice about elimination factors.

Make sure when you consider elimination factors that you are not eliminating people who are different from you *simply because they are different*.

Diversity in an organization takes many forms: Product experience, gender, culture, and race are only a few of the areas where people differ from one another.

In large companies especially, elimination factors may create an unintended but real discriminatory hiring practice. Be sure to ask your corporate lawyer or someone in your Personnel Department or on the Human Resources staff whether your elimination factors might hinder diversity in your workplace. As an example of a potentially discriminatory hiring practice, consider the following: You work for a large multi-site organization. You want to hire a manager to oversee four geographically dispersed sites. You ask the question: "Are you available to travel half of the time?" That's the correct question, but you may well be discriminating against a candidate who could successfully manage the job without traveling at all. Consider whether you can specify the position without the travel requirements to allow people who won't or can't travel to apply for and appropriately fill the position.

Travel can be a problem for non-managers as well. If your organization supplies on-site support to customers in a variety of locations, you may believe you require a customer-support engineer who can travel half of the time. However, by specifying travel as a requirement, you may be ruling out primary-care-givers, physically handicapped people, and more. That can be illegal. Instead of requiring travel, consider setting up alternatives such as videoconferencing, local support staff, self-diagnosing hardware, and so on.

Frequent travel may be a requirement for various categories of technical staff—systems architects, project managers, systems engineers, product managers, and senior-level designers and managers, for example—but make sure to note the reason. If you need architects or designers to travel for a week each quarter to develop the next-generation product line with their peers, that's a different travel requirement than requiring a systems engineer to travel three weeks out of every four to a customer to elicit requirements. If you note why you require travel, you can develop effective interview questions. Then, if a candidate asks for the reasons behind the travel requirement, you can quickly and easily explain the reasons.

Make sure your elimination categories do not exclude handicapped people from your hiring process because of their disability. Discriminating against disabled people in the United States (and most other countries) is illegal, of course, but it is also foolish and shortsighted: A physical handicap most probably will not impede mental performance, and a mental handicap most probably will not impede performance of a primarily physical job.

Now that you've analyzed the job, it is time for you to complete the job analysis worksheet so that you can create a precise and practical job description.

Complete the job analysis worksheet.

Use your worksheet as a place to capture the job analysis work you've done so far. Describe job components, cultural attributes, education, and company factors, and use what you have recorded as a basis for writing a job description.

Once you have filled out the worksheet provided as a sample below, take a look at the case study example following it. The Walker Software Case Study appears in bits and pieces throughout the book and in Appendices A and B.

Defining Questions	Needs & Observations
Who interacts with this person? What roles does this person have in this job? What level is the company willing to pay for? What's the management component?	
What are the job's activities and deliverables? What periodic deliverables are required?	
What are the *essential* qualities, preferences, and non-technical skills? Initiative? Flexibility? Communications skills? Ability to handle projects of varying scope? Ability to work on multiple projects at one time? Influence and negotiation skills? Goal-orientation? Technical leadership and problem-solving skills? Responsibility and independence? Passion for learning? Teamwork skills? Others?	
What are the *desirable* qualities, preferences, and non-technical skills?	
What are the *essential* technical skills? Technology/tool skills? Functional skills? Product-domain skills? Industry experience? Others?	
What are the *desirable* technical skills?	
What minimum level of education, training, or experience is required?	
What are the corporate cultural-fit factors? What benefits should be offered? Company growth? Cash position? Industry leadership? Entrepreneurial environment? Benefits? Company size? Others?	
What elimination factors should be considered? Travel? Availability? Salary? Others?	

Worksheet 2-3: Job Analysis Worksheet (with Job Requisition Name).

Case Study: Walker Software

Now let's walk through the job analysis worksheet using an example from Walker Software, a fictional hundred-person company that manufactures an add-on to the telephone company's central office switch. Walker is coming out of its start-up phase and is becoming profitable, but members of the management team are cautious about growing too rapidly— they have only approved hiring people they can support with the current revenue stream.

Who should we know about on Walker's current staff? First off, Vijay is Director of Engineering and Operations and is responsible for product development and support, with a total staff of thirty people. His direct reports are Dirk, a development manager; Susan, a test manager; Ed, the support manager; and two project managers. Vijay's problem is a simple and common one—he needs more people to do more of the same kind of work.

Dirk manages eighteen people in development, five of whom are senior-level developers and technical leads, eight of whom are mid-level developers, four are junior-level (but not entry-level) developers, and one is a release engineer. Dirk is looking for a junior- to mid-level developer.

In the test group, Susan manages four people now: One is a technical lead/senior-level tester, two are mid-level testers who can read and write code, and one is a junior-level tester. She's looking for one mid-level test automation engineer and a mid-level black-box tester.

Ed has five people in support: Two are tier-1 engineers, two are tier-2 engineers, and one is a tier-3 engineer. Ed is looking for an additional tier-2 engineer.

Also on Vijay's staff are two project managers, but he wants to start up a third project and needs another project manager. The technical people are matrixed into the projects (they report to a functional manager, but perform day-to-day work for a project manager), so the project managers need to be relatively senior, to deal with the issues involved in selecting and negotiating for the appropriate people for their projects.

Dirk's first-draft Software Developer job analysis worksheet is shown below.

Defining Questions	Answers re: Needs & Observations
Who interacts with this person? What roles does this person have in this job? What level can we pay for? What's the management component?	*Works with developers, testers, writers, and project manager as a software developer. Mid-level, no management component.*
What are the job's activities and deliverables?	*High-level design (post architecture), implementation, participate in requirements-definition meetings, moderate and attend design and code reviews, contribute to smoke-test suite, develop integration tests. Able to manage subsystem development under the guidance of a technical lead.*
What are the essential qualities, preferences, and non-technical skills?	*High collaboration, high teamwork, adaptable, able to consider multiple designs.*
What are the desirable qualities, preferences, and skills?	*High focus.*
Essential technical skills	*C++, UNIX system calls, UNIX shell scripts. Data structures. Understanding of telephony industry.*
Desirable technical skills	
What minimum level of educational or training requirement is needed?	*Four years working as a developer, at least two completed projects. BS nice, not necessary. BS equivalent okay.*
What are the corporate cultural-fit factors?	*Small company, project-oriented, high growth expected, stock options.*
What elimination factors should we consider?	*$70,000 cap for salary.*

Worksheet 2-3: Dirk's Job Analysis Worksheet for the Software Developer Opening.

Dirk, Susan, Ed, and Vijay are all looking for someone with at least four years of technical experience. Actually, they're each looking for someone who's completed a variety of smaller projects, and putting the experience level at four years is their shorthand way of saying so. Vijay is looking for someone who's already managed at least two projects from start to end. The qualities, preferences, and skills are similar across group lines, but they're not the same; for all of the job analyses, see Appendix A.

POINTS TO REMEMBER

- Develop a job-analysis method for defining the position's requirements. The interview team will use the requirements to evaluate a candidate's work and cultural fit.
- Consider the context of your group and company when you analyze the job.
- A person's qualities, preferences, and non-technical skills have a huge impact on his or her ability to work successfully in your organization. Understand what's important to the company and to your group.
- Don't make certification or formal education the basis for a job description or for a hiring decision unless your culture requires the degree or you're building a professional services organization.
- When you require the services of a licensed professional, make the license a required component listed in the job analysis.
- Remember your company's attributes when analyzing the job. What appeals to one candidate may not appeal to another, and you want to make sure you attract the candidates to whom your company will best appeal.
- Use elimination factors to eliminate people who won't work out, even if they have all the other technical skills, qualities, and preferences to succeed in your culture. You will be doing yourself *and* the candidate a favor if you address these factors before the candidate accepts the job. Do not use elimination factors to discriminate.
- Record your requirements on a worksheet, so you can refer back to them when you want to hire again.

3

Writing a Job Description

"Hey, Jack, did you see the job description SuperSoft posted? You'd have to have worked twenty years just to obtain the technical skills they want, and they only want a junior person. What planet are they on?"
—Disillusioned job-hunter

The job description for your open position has a specific purpose: It helps you identify candidates with appropriate qualities, preferences, and skills, and it enables you to screen out unsuitable candidates. A well-written job description helps you screen candidates by identifying the technical and non-technical aspects of the job.

If you don't perform the analysis first, you run the risk of missing essential cultural qualities, preferences, skills, or deliverables. Separating the analysis from the description is the same as separating the requirements from design in a project: You develop a different perspective on the problem, and you're freer to iterate on both the analysis and description. Remember that for every two minutes you spend iterating on the job analysis and job description, you probably save yourself at least thirty minutes by not having to phone-screen unqualified candidates; you may even save as much as sixty minutes by not conducting an interview you shouldn't have scheduled.

With a completed job analysis, you have the building blocks for a job description.

The job description is the way the hiring manager communicates the job requirements to the interview team and any recruiters. If you choose to do so, you may also use the job description to communicate job requirements to the candidates.

In addition, you also can use the job description to generate your internal job postings and external advertisements and to guide your interviewers when they develop and ask their interview questions.

Write a clear job description.

In Chapter 2, "Analyzing the Job," you developed a job analysis worksheet to define the job requirements. Now, use those requirements to write a job description. Use a template for the job description such as the one in the sample shown below. (Note that the one in the sample assumes that your audience is either internal or outsiders familiar with your company's culture and hiring tendencies, such as external recruiters.) Categories named in the template are explained in the paragraphs that follow it.

Job Description Template

Job title:

Reporting-to manager's title:

Generic requirements:

Specific requirements:

Responsibilities:

Elimination factors:

Other factors:

Job title: Identify the job by title, being as precise as possible, especially if departmental peculiarities must be taken into consideration when filling the position.

Reporting-to manager's title: Specify to whom the job holder will report, by name and by title. Being specific to this degree will prevent people from confusing your job opening with someone else's.

Generic requirements: List the minimum level of education or training required for this job level at this company. Do not bother describing levels that are industry-specific but are not relevant to the opening you wish to fill. By knowing the minimum generic requirements, you and your recruiters can more easily screen out unqualified candidates. List requirements you would expect anyone in the field to have, but be sure they are compatible with the essential technical skills identified in your job analysis. An example of a generic requirement might be stated as "two years of experience in a mainstream software language such as C++."

Specific requirements: Based on your job analysis, which indicated who would interact with this person and the essential technical skills, list any other desired qualifications here. Use the specific requirements to differentiate the specific knowledge you need in your particular company from the generic requirements for the position. Derive the specifics from the cultural qualities, preferences, activities, deliverables, and skills identified in the job analysis. An example of a specific requirement might be stated as "two years of machine-vision experience."

Responsibilities: You will need to record the essential job activities, deliverables, cultural qualities, preferences, and skills desired in the position, but the responsibilities section of the template allows you to further refine the requirements to the specific environment. Responsibilities first can be defined by the deliverables in the job analysis; then modified by the essential qualities, preferences, and skills; and finally, the description can be made even more precise by addition of details about desired characteristics, temperament, or personality. For example, you may want a project manager who can get the best effort out of shy-but-talented developers. You may well include "manage projects" as part of the responsibilities requirements, but your awareness of the people already working on the project would merit your also noting something such as, "Must manage projects staffed by shy-but-talented developers."

If the project manager must deal with developers who aren't known for their teamwork abilities, you could say, "Must manage projects, using significant diplomacy with highly skilled, independent-thinking developers who care about the project above all else." Choose your wording carefully while keeping in mind your job description's potential audience. Some of these talented developers may see the job posting or participate on the interview team, and you want them to be happy with the description. Choose your words so as to attract potentially successful candidates without alienating the interview team.

When describing the responsibilities, avoid using jargon. Write short, clear descriptions of the responsibilities. Even if your audience for the job description is internal, your HR staff may not understand your words. Also, if you choose to provide candidates with a job description, jargon may throw them off.

Elimination factors: Use the elimination factors already noted on the job analysis worksheet. Don't forget to consider salary as an elimination factor, but don't list it unless your company practices open-book management. And, if corporate activities, travel, specific-time availability, or a clear, articulate speaking voice are part of the elimination factors, then say so.

Other factors: An optional bit of information to enter on the template pertains to the corporate cultural-fit factors.[1] For a written job description, whether you include cultural-fit factors will depend on what those factors are and whether you want other people to see them. The corporate-fit factors are critically important because they will impact your ability to screen candidates, so discuss them with any recruiters you use, even if you don't write them down. Unfortunately, those factors may be the most controversial inside your organization. Think carefully about which factors you can write down on a document that will be circulated inside your organization.

A True Story

Stu, a VP in a major insurance company, wanted to describe his organization in the following terms: "Bureaucratic organization moves slowly. Change agents need patience." Instead, so as not to offend anyone within the organization by calling it bureaucratic, he called it "a traditional-thinking organization." That more diplomatic wording enabled him to be able to ask appropriate questions and not offend others in his division who read the description.

The Lesson: If you work for a company whose management discourages you from including corporate cultural-fit factors in a job description, don't include them. But, *do* use those fit factors when evaluating candidates.

Use job descriptions to help you screen candidates.

Clearly, job descriptions generally are useful as early screening devices, but not all job descriptions are worth their weight in gold. I've encountered three kinds of job descriptions that are woefully inadequate:

- *laundry lists*—these endlessly detail everything a person would possibly need to know and do in order to succeed at the most senior level for this job.
- *vague, ambiguous, hand-waving descriptions*—these hint at something that might be a job someone wants done, and ignore the personal qualities, preferences, skills, significant activities, and deliverables.

---

[1] The topic of screening candidates for corporate fit is admirably addressed in Jim Harris and Joan Brannick, *Finding & Keeping Great Employees* (New York: AMA Publications, 1999), pp. 16ff.

- *boilerplate, generic descriptions*—these ignore all the personal qualities, preferences, and skills and assume all people with the title are the same and perform the same work.

A True Story

Following is a verbatim laundry-list job description for a six-month, contract-status, development job. Note that it suffers from the second affliction in the preceding bullet points: It's vague, and it's badly written.

List of Job-Description Attributes	Interviewer's Notes
Software Development: Primarily coding in Java. **Development Environment:** Java, Tomcat. **Web Services:** SOAP, WSDL, XML, Oracle database, PVCS software configuration, management, existing code base is in Visual Basic running on Win 2000 platform.	*Okay, they want Java even though current code is in Visual Basic. Maybe there's a reason farther down.*
Systems Understanding: Ability to understand requirements documents and high-level architecture. Ability to understand legacy software. **Design/Documentation:** Ability to generate software design specs. Ability to generate design to support new requirements on an existing system. **Programming:** Excellent programming skills in Java server-side development. Skilled in database interactions, specifically Oracle. **Web Services:** SOAP—Simple Object Access Protocol. Generate and work with XML schemas. Web Services Definition Language—WSDL. Skills in http data transmission and communications protocols libraries like SAP and socket level programming, and ftp o. **Process:** History of successful software commercialization projects. Knowledge and proven execution of Software Development processes (e.g. SCM). **Quality:** Track record of quality software development including the creation and execution of unit tests, and release documentation to Software Configuration Management. **Environment:** Demonstrated successful projects involving development and integration across multiple development organizations located in different geographic locations. Ability to work flexible hours to accommodate multi-national development. **Education:** BS Computer Science + 5 yrs experience.	*Ability to understand requirements and high-level architecture is fine. What don't the current people understand?* *Hold on. Why XML schemas? How big is this job?* *Design specs are good. If the new requirements are for the existing system, who's porting from Visual Basic to Java?* *There's a lot more to process than configuration management. Is this job the same as that of release engineer?* *The following points stand out for me: Not everyone is in the same location; the work involves mega-hours; position responsible for design, architecture, and release management, maybe porting. All with five years of experience? For a six-month contract position? Uh huh. No one with ten years of experience would touch this job.*

> We should add to the laundry-list job description one more requirement: "Able to leap tall buildings in a single bound." Notice that in this real example, which was posted on a job board I saw during a recent training session, there is no mention of deliverables or activities—that is, there is no description of what the person would actually do.

In case you haven't come to this conclusion from what you have just read, let me state my advice unequivocally: Avoid laundry-list job descriptions. Finding someone who meets all those criteria is impossible. Even if you found someone who met the technical criteria, he or she probably wouldn't have your preferred cultural qualities or meet your salary requirements. Unless you're looking for and are willing to wait for and pay a specific, senior-level technical person, a laundry list of technical skills is unrealistic, and only serves to screen out potential candidates. If you've found that you created a laundry-list job description, because you believe that only a superstar can perform the work, ask for help analyzing the work.

Ambiguous, hand-waving job descriptions don't work either—the more ambiguous the job description, the less pre-interview screening you can do. In the laundry-list example printed above, "Process" and "Quality" are part of the description. What about them? What should a developer know about process and quality? How to spell the two words? How to use process and quality to develop more effectively? Which process? Which quality? The job description is too vague to enable a candidate to know whether he or she measures up.

If you have generic job descriptions, even boilerplates, start with them, but don't use them as they are. Customize them with the qualities, preferences, non-technical skills, and specific technical skills you require, so you can more thoroughly screen candidates. You can't easily or successfully recruit, screen résumés, or interview candidates if you use ambiguous job descriptions.

Job descriptions that only address technical skills without any of the cultural qualities are likely to lead you to interview and hire unsuitable or unqualified candidates. If you include typical responsibilities (deliverables and activities) in your job descriptions, you're much more likely to attract candidates with appropriate cultural qualities as well as with needed technical skills.

Identify who will use your job description.

A job description is like any other work document, with the following factors to be considered:

- What audience or audiences are you trying to reach?
- What level of detail do the various audiences need in the job description?

I write job descriptions for internal use only. I don't post job descriptions; I post ads. For me, "internal use" includes external contract recruiters. Because I develop long-term relationships with external recruiters, I treat them the same way I treat the internal interviewers with whom I work. You can get away with using shorthand inside the company and with recruiters who know you and your company well, but be as specific as possible when writing the description. Determine, for example, whether a recruiter needs more or less information than a member of your interview team.

Work hard to specify why you want someone from a particular industry or with technical experience. The more specific you are in stating your requirements, the more successful your screening will be. If you use shorthand such as, "Five years of experience in the medical device industry," you haven't specified your requirements. That experience could mean experience conducting audits, working with patent lawyers, or devising long test cycles, or it could mean something completely different—and it could mean something different to each candidate as well as to you. Be as specific as you can be in the job description so that you attract the people you need.

Instead of using shorthand, try to express your negotiable tradeoffs in order to attract qualified candidates—such as by specifying "Industry experience with 21 CFR Part 11 or experience in other process-audit methods," for example. The more ambiguous you are, the more time you will have to spend working closely with your recruiters, so that they understand what your shorthand means.

You may be working with an HR representative who doesn't know how best to describe technical work, or who doesn't understand your requirements. In that case, your HR rep may want to use some shorthand technique to describe what he or she thinks the requirement means. For example, HR representatives frequently indicate that candidates need a technical degree, because they don't know how to evaluate functional skill and experience. To eliminate this ambiguity, you could provide the representative with a specific description of the requisite qualities, preferences, and non-technical skills *as part of the job description.* Or, you could do what I generally choose—teach the HR rep how to read a résumé with an eye toward detecting requirements.

On the other hand, if your HR rep is bound by some company policy, or honestly doesn't understand how a technical person could be successful without a degree or without experience with a particular operating system, then choose how much energy you want to spend explaining your choices to your HR rep. Don't waste your time, and don't let the HR rep prevent you from finding the best-suited, qualified candidates.

If you work with external recruiters, it's worth taking the time to discuss your personal as well as your organizational assumptions about title, salary, and job responsibilities, so that the recruiters understand what you mean when you talk about a particular job title. Contrary to what many of us believe, job titles are neither standardized nor meaningful from one company to another—they never mean quite the same thing in any two companies. When I use contract recruiters, I explain our job ladder, salary ranges, and how the company works, so they can find appropriate candidates. If you don't develop long-term relationships with recruiters, you'll have to educate each recruiter about your company every time you use one.

For external posting, I write a separate advertisement or use full sentences to convert the bullets in my job description. That way, I minimize the chance that the reader will make incorrect assumptions about what I mean in a job description.

Consider whether your bulleted job description is appropriately worded if you're dealing with someone who wants to publish it for recruiting purposes. It's too easy to jot down vague descriptions or to make an assumption that the reader will understand what you meant (even if it isn't what you wrote) when writing the first draft of a job description.

If the people who will be using the job description are recruiters or interviewers, make sure they know why you're looking for the specified kinds of experience. Use the job description as a starting point for your conversation about this open job.

Learn how best to use standardized job descriptions.

Some organizations do use standardized job descriptions to describe each role on their technical staff: developers, testers, writers, and so on. Standardized job descriptions are a great way to identify what's common among your technical staff members. However, unless you are unique among technical organizations and have fungible staff members who can easily replace each other, you'll need to augment the standardized description with your needs for this particular position.[2] Use your job analysis worksheet to differentiate the candidates you want to attract from those matching the standardized job description. Then, add your changes.

You don't have to use my job description template, especially if your company has its own template. Just be sure you have a place to map all the parts of the job analysis to whatever job description template you use. If you don't, you run the risk of not being able to screen résumés carefully enough, and you'll waste time phone-screening or even interviewing people who are not suitable candidates.

Develop your job description over several drafts.

Don't expect to write a perfect job description the first time. As with any writing, you'll find it easier to write a first draft quickly, put it aside, and modify it later. If you're pressed for time, ask people in your group to review what you have written—and expect to modify and improve the job description while you interview.

As you write ads or phone-screen scripts, you may remember something you want to change in the job description. That's fine—change it. When I start with a new job analysis, I plan on creating two or three drafts of a job description. After I've phone-screened or interviewed a few candidates, I may have more changes for the job description.

If you're having trouble writing the job description, ask the rest of your group, interview team, or the candidate's would-be customers—the people with whom the employee will work—to help you. Your team or customers may be better able to articulate what you're looking for because they are living with the lack of adequate staff every day. You don't have to write a job description by yourself. If you think you could produce a job description that would be more specific or more accurate if you had other people's help, write that job description as part of a team effort. Not only will working with others help you, your interview-team members will have a

---

[2] The myth of fungible resources is exposed brilliantly in Tom DeMarco, *Slack: Getting Past Burnout, Busywork, and the Myth of Total Efficiency* (New York: Broadway Books, 2001), pp. 12ff.

better idea of what questions to ask in the interview. And, you'll have a chance to start mentoring your team, in case members of it ever become hiring managers.

Case Study: Walker Software

Dirk Jones has prepared a draft Software Developer job analysis worksheet. Although he is concerned he's spending too much time planning and not enough time interviewing, he's serious about finding the right person quickly, and he's willing to give the job-description exercise a try. At first glance, his template looks fairly complete. His second-draft job analysis worksheet appears below. Immediately following it is Dirk's fleshed-out job description (my comments are noted in italics to show what the description tells me).

Defining Questions	Answers re: Needs & Observations
Interaction, roles, level, management component?	Works with developers, testers, writers, project manager, and product manager (or customer representatives) as a software developer. Mid-level, no management component.
Deliverables and activities?	Deliverables: high-level design (post architecture), implementation. Activities: participates in requirements definition meetings, moderates and attends design and code reviews, contributes to smoke test suite, develops integration tests. Leads subsystem development under the guidance of a technical lead.
Essential qualities, preferences, skills?	High collaboration, high teamwork, adaptable, able to consider multiple designs.
Desirable qualities, preferences, skills?	High focus.
Essential technical skills?	C++, UNIX system calls, UNIX shell scripts. Data structures Understanding of telephone industry.
Desirable technical skills?	
Minimum education or training requirements?	Four years working as a developer, at least two completed projects. BS nice, not necessary. BS equivalent okay.
Corporate cultural-fit factors?	Small company, project-oriented, high growth expected, stock options.
Elimination factors?	Salary no higher than mid-range developer.

Software Developer, reporting to Development Manager
Generic requirements:

- Minimum of 4 years as a developer, on at least 2 completed projects
- 2 years of C++
- 2 years of UNIX system calls, data structures
- BS, CS, or equivalent experience. *(Dirk doesn't care whether his candidates have a degree.)*

Specific requirements:

- Understanding of telephony industry. *(Dirk hasn't yet explained why this is necessary or what part of the industry the candidate should understand.)*
- At least 2 years of work on the XYZ subsystem, with responsibility for implementation and unit-testing on completed projects. *(Dirk is looking for someone who's already learned how to design, and who now is learning how to develop the system's architecture. He wants someone who understands how to maintain and add onto code that may not have been sufficiently designed or tested. Note that Dirk hasn't asked for what he wants; he's using this explanation of experience as shorthand for what he really wants. For a first-draft job description, this is okay.)*

Responsibilities:

- For a subsystem: Assist with requirements and overall architecture; design, implement, and unit-test the subsystem
- Moderate and attend design reviews
- Moderate and attend code inspections
- Implement smoke tests for the subsystem
- Develop integration tests for the subsystem
- Work with developers and testers to design, develop, and unit-test the subproject. *(This addresses the high collaboration and high teamwork parts of the essential qualities.)*
- Develop multiple designs for a specific problem. *(Again, part of the essential qualities.)*
- Design for performance and reliability. *(This addresses why Dirk wants someone with telephone-industry experience, so the person already understands the implicit requirements. This extra responsibility wasn't already specified, but was derived from the telephone-industry experience.)*

Additional responsibilities for the appropriate candidate *(Use desirable skills, qualities, and preferences.)*:

- Able to focus on own tasks, even when the rest of the group is working on other tasks. *(Dirk has had problems in the past with developers who couldn't stay focused on their own work, but wanted to solve other people's problems. He's not sure this is the correct way to ask for a person with "high focus" who will mind his or her own business and keep working, but he figures it couldn't hurt.)*

Other factors: None. *(Dirk does not publish salary as an elimination factor. He will find out about salary during the phone-screen.)*

Template 3-1: Dirk's Software Developer Job Description.

Dirk isn't thrilled with how he's described his need for someone with "high focus," but he's written it, and he's clarified why telephone-industry experience is important to him. He can refine the high-focus part later. His job description is still vague as a result of the "understanding of telephone-industry" requirement. In reality, how any candidate understands performance and reliability criteria is critical to product success—and the experience category is appropriate—but Dirk isn't thinking about other industries. That's okay, he'll consider how to ask for what he wants when he writes a phone-screen and develops interview questions.

Now that Dirk has prepared the job analysis worksheet and job description, he can write job postings as well as advertisements.

POINTS TO REMEMBER

- Create a job description that's as specific as you can write it for the position you're trying to fill. If you are trying to fill more than one position of the same kind, list the tradeoffs in the desirable requirements area of the job description template.
- Consider who will use the job description, and write the job description so that person can successfully use it.
- As you write the job description, check to see that you've mapped all the pieces of your analysis to the job description. Decide whether you want to include the elimination factors or company factors as part of the job description, even if the description is just for internal publication.
- Expect to refine, expand, or modify your job analysis when generating the job description. The more you think about a particular problem (that is, the more you think about the reason you're hiring someone), the more you'll understand about how to hire to best fill the opening.

Part 2

Sourcing and Selecting Candidates to Interview

You've defined your strategy, analyzed the job, and created the job description. Now it's time to choose how to source candidates, so that you can write advertisements and review résumés.

If you'd like to measure how well you're doing, use the following three steps to measure your various recruiting strategies:

1. Build a table with five columns, as follows:

Recruiting strategy	# of résumés received from this strategy	# of phone-screens	# of in-person interviews	# of hires

2. As you use a new recruiting strategy and receive résumés, add that strategy to the table.
3. Review your data every month. When you must hire a large quantity of people, review your numbers even more often, possibly biweekly or weekly. Look for sourcing strategies that don't just bring in numerous résumés, but for strategies that pay off in number of hires. Which strategies are working for you in terms of total number of hires? Which strategies take too much of your time for too little return (number of total hires)? If your sourcing strategies are not working, review your job analysis and job description sheets—they may not be specific enough, or they may not be attracting the people you need to attract.

Each sourcing strategy will produce more or less success based on the economy, the job, and your company's reputation. Since you'll be trading off your time against your company's money, it's worth tracking how valuable each recruiting strategy is to you.

The three chapters in Part 2 detail how to recruit and screen candidates from multiple sources.

4

Sourcing Candidates

"Okay, we put the ad on our Website. We can just sit back and wait for the résumés, right?" —Optimistic hiring novice

"Sourcing" is a word used by recruiters to mean "attracting" suitable candidates who can become productive employees. Effective sourcing techniques usually require a large component of either time or money, but by specifying and developing a range of sourcing techniques—the Web as well as local and trade newspapers, for example—for use during the recruiting process, you can exploit your sources to attract the most suitable candidates. The sourcing techniques presented in Table 4-1 reflect ways to use time and money to your best advantage.

Sourcing techniques that can require more time than money	Sourcing techniques that can require more money than time
Corporate Website ads	Newspaper ads
Employee referrals *(in some circumstances)*	Radio ads
Job fairs	Internet ads
Local professional groups	Employee referrals *(in some circumstances)*
Professional group meetings	Internal recruiters *(in some circumstances)*
Conferences	External recruiters
Customer networks	Executive and retained-search recruiters and headhunters
Personal networks	Other nontraditional ads
Candidate referrals	
Interns and co-ops	
University affiliations	
College recruiting programs	
Web-based résumé boards	
Internal recruiters *(in some circumstances)*	
Former employees	

Table 4-1: Time/Money Sourcing Techniques.

Use time, not money, to attract suitable candidates.

You don't necessarily have to spend carloads of money to find great candidates, but if you don't, you probably will need to spend a good bit of time. Some techniques to try if you must dedicate more time than money to the recruitment effort are detailed below. Remember, potential candidates may not all look in one place to learn about the great job you have open, so you will want to use a variety of sourcing techniques to reach them.

Corporate Website ads

Post all your open positions on your corporate Website. Use the following checklist to make sure the Website is working for you:

- Is your job-openings page easy to find?
- Is the job-openings page just one click from your homepage?
- Is the ad for your open position on the job-openings page?
- Does your search engine index the content of the ad?

If you can highlight your job on your company's homepage, do so. If you can't obtain space on the homepage to feature your open job, see whether the Webmaster will highlight the job-openings page in some way.

If you're advertising a hard-to-fill position, make sure a description of the position is highlighted on your Website. If your company has many open positions, make it easy for people to find positions by either geographic location or type of position. Beware of industry-specific or company-specific job titles or terms—you may lose someone who could do the job, but didn't know what to look for.

A True Story

One hiring manager from the pharmaceutical industry was desperately looking for a tester. Unfortunately, his company's job page described everyone affiliated with software as a "business analyst," whether the person was a business analyst, a developer, a tester, or a project manager.

If you encounter this type of naming problem, explain to the people who manage Website content that visitors to the site may not realize that a "business analyst" may well be what they would call a software tester or developer.

Employee referrals

Your employees may give you the best candidate referrals you'll ever get from anyone. After all, who has a better understanding of your company and its culture than your current employees? Your employees are also motivated to attract people with whom they can work easily.

Employee referrals are part of your *continuous recruiting strategy*, whether you have open positions or not. If someone says, "I know of a great person for job X," but there is no opening, be sure you say, "We don't have an opening right now, but please send me your contact's résumé." When you do have an opening for job X, you'll already have a list of candidates. Later, when you have an open position and a recruiter sends you the same candidate, explain that you already knew about the candidate, and you won't need to pay the recruiter for that candidate.

Employee referrals allow members of a department, division, or team to be an integral part of the hiring process from the beginning, a good technique for integrating the eventual employee into the group. How many employee referrals you receive is a good barometer for you and your company, because people will not make a referral if they don't like working for you—or for the company. If employees provide lots of referrals, it probably means they enjoy their environment.

One way to encourage referrals even if the environment is not all peaches and cream is to pay for referrals. I recommend that you pay your employees for each referral in the same way you would pay an external recruiter, observing the same waiting period you observe before paying a recruiter for a placement. Use your judgment about how much to pay for referrals. If the payment is too high, members of your staff may want to start recruiting instead of doing their technical work. If the payment is too low, they may not bother bringing you candidates. Make the referral payment sufficiently valuable so that your staff will bring candidates into your company.

Job fairs

Attend local, well-advertised job fairs. A fair that is held near your place of business will help keep travel expenses down and will eliminate the need for hotel accommodations; a well-advertised fair is likely to draw a sizable, qualified pool of candidates. Depending on the size of the fair, plan to reserve a table or a booth from which you can work. Bring at least a couple of other hiring managers or internal recruiters with you to the fair so that all of you can work in parallel, asking candidates questions. Prepare for the job fair by putting together colorful flyers as ads for your open positions, and develop three-to-five, close-ended and three-to-five, open-ended questions to ask of people who stop at your table (see Chapter 7, "Developing Interview Questions and Techniques," for an in-depth discussion of suitable question types). When you meet candidates at the job fair, quickly screen them by asking these questions to decide whether you're interested in following up with a phone-screen or an interview, or to recommend the candidate to one of your peer hiring managers. Be ready with a schedule of available phone-screen or interview times, so that if you find a "hot" candidate, you can set the follow-up time immediately.

When you bring others with you to the job fair, review the requisite social skills for success. Each person doesn't have to go out of his or her way to make small talk, but everyone does need to be able to attract a candidate to your booth, to engage a candidate in conversation long enough to assess basic character and personality traits, and then to determine whether the candidate has the minimum skill-set for the job.

I've found that the following technique works well at job fairs: I position myself near the front of the booth and smile whenever I make eye contact with people as they approach the booth, so as to look welcoming but not overly eager or intimidating. I never stand at the back of the booth or talk to a coworker if I can possibly do otherwise, as both of these stances make it look as if what I am doing is more important than talking to potential applicants. It has been my observation that my chances of attracting candidates into the booth decrease the farther away from the front of the booth I stand. If you opt to do nothing else I have suggested, the following is a must: Avoid scowling, and avoid carrying on long conversations on your cell phone, even if someone has just called with bad news about your current project.

When a person comes up and looks interested, introduce yourself and ask whether he or she has questions about the company or the job(s). If the individual has questions, answer them, and then ask for a copy of his or her résumé. If the person has no questions, briefly describe what type of business your company does or some of the jobs you hope to fill, and then ask for the person's résumé. Take a quick look at the résumé, and then ask your set of six-to-ten questions to determine whether the candidate has the minimum skill-set. If the individual does have the requisite skill-set, try to establish a time for a follow-up; if he or she does not have the minimum skill-set, thank the person for stopping by, and acknowledge that although the person probably is not right for the particular job, you'll keep the résumé on file or, if you know of another company or job the candidate fits, you can suggest that the candidate might investigate it (at the job fair or not).

If you have a coworker with you, you may transfer the candidate to the coworker for the open-ended questions. No matter who asks the open-ended questions, if you like the answers, direct the candidate to the person from your company who has taken responsibility for the phone-screen or interview time sheet. An appointment for a follow-up phone-screen or an interview should be made on the spot.

This whole process should take about ten minutes per candidate. You don't want to spend a lot of time with any given candidate, because you want to make time for as many people as possible. Similarly, candidates want to be able to move on to other booths, and will feel ten minutes is a sufficient amount of time to spend with you. Keep in mind that the purpose of a job fair is to enable you to perform initial screening on the spot, but it is also designed to give potential candidates time enough to assess you and your company but not bog down in minutia.

Make curiosity your assistant at a job fair. Work to make each candidate curious about your company, your group, and the work you need done. To make candidates curious, first ask enough of your close-ended questions—such as the elimination factor questions in your phone-screen (see Chapter 8, "Creating and Using Phone-Screens")—and then ask some behavior-description questions to verify that the candidate has enough experience to justify bringing him or her in for an in-person interview. Tell the candidate about your main attractor (see Chapter 5, "Developing Ads for Open Positions").

Use your behavior-description questions to find out about candidates, and to challenge them a little so they're interested in coming back to you for a real interview.

If you have more people staffing the booth than just one or two others in addition to yourself, you may be able to set up a kind of reception line, moving suitable candidates from booth member to booth member, each of whom can be assigned one or two questions to ask each candidate. I call this reception line a "question line." Once the candidate has answered a question, he or she then moves to the next booth member in line. Although each questioner sees a less detailed picture of the candidates than would be seen if he or she were personally asking all questions, more people can form initial opinions and first impressions, usually a good enough indicator of whether the candidate should be pursued further.

Make the question line fun, possibly almost like a game-show event, but take care that people are not made to feel intimidated or to think they might be the brunt of a joke only you and the booth staff understand. This is a serious danger: Question lines *can* intimidate candidates. I've seen some lines set up so that a candidate can be ejected partway through. That kind of treatment is not only disrespectful, bordering on cruel behavior, it also shows hiring staff members in the worst possible light. I'd rather spend more time with each candidate, even the ones who ultimately will not get called back for an interview, than too little time; and I'd also rather conduct preliminary screening in-person, possibly only asking three questions, and then use a phone-screen later. Another way I've found to be effective to handle this situation is to bring more people from my hiring team to the job fair. That way, other people besides myself are available to engage and interact with candidates before they walk by.

Local professional groups

Learn which groups have regular meetings in your area. Professional and user groups with a presence in most urban areas of the United States are listed in the following table, but similar groups exist throughout the world. If you are interested in finding what's nearest you, consult telephone listings and academic directories for your area.

Acronym	Full Name of Organization	Applicable Positions
SPIN	Software Process Improvement Network	QA, testing, process improvement
ASQ	American Society for Quality	QA, testing, process improvement
IEEE	Inst. of Electrical & Electronics Engineers	Developers
ACM	Association for Computing Machinery	Developers
PMI	Project Management Institute	Project managers
HDI	Help-Desk Institute	Help-desk managers
STC	Society for Technical Communications	Writers

Table 4-2: Notable Professional Groups.

Most of these professional societies have divisions or special interest groups. And most have local or regional groups. Don't forget user groups for tools you use, such as UNIX, Java, Oracle, ClearCase, and so on.

There may be more local groups that aren't part of a national organization, so look for information about these groups and attend meetings. If you frequent these meetings and let people know that you are a hiring manager or a member of a hiring team, people will learn to seek you out to see whether you're hiring.

Just as I've recommended you do prior to participating in job fairs, prepare bright-colored flyers that will inform prospective candidates about job openings. Bring plenty of flyers to each meeting, but be sure to ask the specific group's organizers or the association's directors whether you can announce that you have open positions and distribute the flyer. Don't distribute material without first getting permission to do so. The reason for this should be obvious, but I'll note it anyway: You want to be able to come back to subsequent meetings as a welcome participant, and that can only happen if you show the organizers that you recognize that you are their invited "guest."

In addition to opening their meetings to you, some professional groups may make lists of their members' e-mail addresses available to you or may allow you to advertise your open positions by means of a link to your Website. If you are permitted to take action in either of these areas, be sure to learn precisely how the groups will want you to post your open jobs or to e-mail their members. If you're sending e-mail to a professional group, briefly describe open jobs in the text of the e-mail and supply a URL to link to the ad on your Website; do not include an attachment. Not everyone can—or will—open an attachment from someone whose name or affiliation is unknown.

If you're looking for a large number of people to hire, or for a specialized, senior-level person, consider hosting a professional-group meeting. Unless you appoint someone to handle logistics for you, as the host of the meeting you'll have some headaches—such as having to arrange for a suitable place to meet, speakers, audiovisual equipment, room furnishings, refreshments, and the like—but you'll also have unrestricted access to people interested in the topic who, presumably, are also skilled professionals. People who attend or speak at professional-group meetings may be perfect candidates for a job you have open. They may or may not be looking for a job, but you'll have first crack at them if they find that what you have to offer fits the bill.

Conferences

If you haven't attended a professional conference in a while, it's time. Conferences provide you with an excellent opportunity to meet people with similar interests. Participation in professional conferences can help you build rapport with hiring prospects without having to conduct a formal interview. Most conferences have a designated message board or an area where announcements can be posted—you can post job openings, and let people know how to contact you. You can also network with other hiring managers to learn about their hiring strategies.

Participation in local and regional conferences is an especially good way for you to meet people who don't require relocation reimbursement. You can also use local people as part of your professional network, to help spread the message that you're hiring. If you're trying to hire a sizable number of people and the conference offers an exhibitors' section, consider paying for a booth, and staffing the booth during conference breaks.

National or international conferences can also help you with sourcing efforts. Just be sure to make it clear to conference attendees whether or not you're willing to reimburse them for relocation expenses.

Customer networks

In my experience, other people are happy to network and suggest candidates to you. Let your customers know you're looking for people to hire. Sometimes, your customers may enjoy working with your product so much they'd like to be involved in its maintenance. Alternatively, they may know people who'd like to work on your product. If your customers want to work for you, clarify both your job criteria and your policy about hiring customers. I've met some salespeople who thought it was okay to promise customers jobs if they would buy the company's equipment.

Personal networks

Your personal network is valuable. Do you try to meet people at professional group meetings, conferences, and even vendor user-group meetings? If not, you're missing out on a great way to find potential candidates. Consider bringing to every professional-group meeting you attend promotional material that points to your company's Website or job page, so people who want to learn more about opportunities at your company can do so.

Use your personal network wisely. You don't have to deluge your network with e-mails every time you're looking for a candidate, but let key people know about openings about once a quarter.

If you are having trouble finding appropriate candidates, don't forget that your friends and relatives can be excellent resources. Sometimes, the art history major you roomed with in college or your Cousin Vinnie's nephew Bobby knows exactly the right person for that project management job.

The more hiring you do, the more worthwhile it is to work to increase your network. Similarly, the more varied your network, the more likely you are to be able to find suitable people.

Candidate referrals

Sometimes, when you phone-screen a candidate, you realize that he or she is not right for *your* specific open position, but that the person could be perfect for a position you know of at another company. In general terms, tell the candidate both what makes him or her not right for the specific position you are trying to fill, and also what you believe makes him or her suitable for the other company. Offer to give the candidate contact information at the other company if he or she seems receptive to the idea—but only if you truly believe the person stands a good chance of being hired. Do *not* send a candidate on a wild goose chase just to get him or her out of your office! And, do *not* give too much detail about why the candidate is not qualified for the job you have open as that can lead to a verbal shooting match in which the candidate feels compelled to argue each point.

Assuming that you have established a good rapport with the candidate you are offering to refer elsewhere, you might ask whether he or she can return the favor by recommending a colleague who might be suitable for your opening. This request for a favor from someone who is essentially a stranger and who may feel some sense of anger at being rejected (despite your assurances to the contrary) must be handled very, very carefully. If you are confident that you can pull this information exchange off, go ahead and request the favor; you may find that the candidate will tell you about colleagues who are just what you are looking for.

Interns and co-ops from local colleges

I've had great results hiring college students, both as interns and co-ops. For readers who are not familiar with such programs, a co-op enables students to alternate taking academic courses certain semesters with working part- or full-time during other semesters. Interns also are students, but they may or may not get academic credit for the time they spend working, depending on the requirements of their specific college or university, and sometimes even of their major course of study. Interns generally are available to a single company for one semester only whereas co-op students might be permitted to work for you each semester their school has designated as a work semester, thereby returning more benefit to you for your training effort. Participating universities generally help place undergraduates with you, and most favor establishing corporate affiliations.

The reason I am so in favor of intern and co-op programs is that they provide me with a way of sampling student caliber at local colleges and universities to see whether undergraduates at these schools can apply the relevant technical and functional skills in my positions, and the program provides me with an opportunity to hire the now-familiar students as full-time staff when they graduate.

Of course, you'll want to keep in mind that these people are undergraduates, and may not have the emotional maturity or technical experience you would expect of a college graduate. However, I've found great people for no referral fee and a comparatively small salary. Many students make excellent full-time hires when they graduate: They're not tainted by a bad experience working elsewhere, they're enthusiastic, and they've already started to master the learning curve about your product domain. In addition, they are a known commodity, reducing your risk of discovering you've made an unsuitable hire three or six months into a new employee's tenure.

University affiliations

Once you're ready to move beyond the intern or co-op route and you have some clout in your company, find a university professor or group of professors doing work you want to be affiliated with, and support their research, either through direct funding, product contributions, or the like. For example, if you're in the telecommunications business, you could support research in fiber optic communications or in increasing communications bandwidth. If domain expertise isn't as important as general technical skills, then identify a cadre of university professors who give their students test-case projects that you like, and fund their research.

Talk to any people in your company who fund research and development, and ask about funding an R&D grant for a professor or the students. If you establish a strong enough connection, you can get access to the best graduate students and also to professors who will steer their top graduates to you.

College recruiting programs

A university affiliation that differs from the one described above is what I call a university-based, new-hire program. If you build a relationship with the schools from which you seem to hire most often, they may be willing to arrange on-campus interviews with seniors mid-way through their final year. This way, you get the chance to meet with interested seniors, and you can make offers to those you wish to hire. Obviously, such undergraduates usually won't be able to start work until they've graduated, typically in May or June but sometimes as early as December or January, but you will have spent less time and considerably less money than if you had used an external recruiter, making the wait worthwhile. In some cases, seniors will be available to work on a part-time basis even prior to graduation, so getting in contact with them early on can be beneficial to them and to you.

Depending on how formal your company is or how structured the university's placement and alumni offices are, you may need to have your Personnel or HR Department set up this kind of recruiting visit for you, but do whatever you can to make the most of college recruiting.

There is one thing to watch out for with recent college graduates, however: You may have to teach them *how to work* as well as *how to do the work*. If they have never held a professional-level job before, you may have to explain more about protocol and work ethic than you would to a new hire who is seasoned. It's generally worth it, however, because you'll end up with a well-trained staff member who has not been exposed to someone else's bad habits or bad management practices. The drawback can be that he or she may take longer to become productive than someone who already knows how to settle into a job.

Web-based résumé boards

Take advantage of Web-based résumé boards, such as Monster.com, Techies.com, Dice.com, and others. Although you'll probably have to pay to post an ad, the rates are generally cheaper than traditional print advertising. Look for local Web-based résumé boards that focus on your geographical area. Geographically centered résumé boards, such as Bostonworks.com for people like me who are based in New England, are generally hosted by local newspapers. As I noted above in the discussion about conference participation, specify whether you will pay relocation expenses when you place an ad on a geographically specific Website.

The bigger Web-based résumé boards can also help your screening process because their database filter will analyze and digest résumés and partition them into keyword categories.

Former employees

If your company has been in business for more than a few years, you may have "alumni"—people who previously worked at your company but no longer do. Some of these people may have left your company for reasons that had nothing to do with the work (leaving, for example, because a spouse was transferred or to take time off for a baby), and are now available to work for you again. Determine if there is a mailing list of previous employees, and think carefully about whether you want to recruit from that list. Previous employees can bring a significant advantage to the company over people who've never worked for the company: They have domain expertise in your products and a proven ability to learn the products, for example. Make sure you evaluate each person's candidacy in light of his or her reasons for leaving, and speak with previous managers or peers before you contact the former employee, to make sure the person truly is suitable as a candidate.[1]

Develop a continuous recruiting program.

Your recruitment process does not have to end when a position has been filled. Consider continuous recruiting, especially if you work for a company that can be more flexible about creating open positions.[2] Or, if you work for a company that is growing aggressively, but which only opens positions at certain times of the year, consider continuous recruiting. If you work for a small company and you're looking for specialized expertise at a senior level, where it's difficult to train people, then it might make sense for you to look continuously for great candidates. Leave the job description up on your Website, let your internal and external recruiters know, and tell your network of colleagues that you're always looking for this particular kind of person.

If you're in continuous recruiting mode, choose which of the techniques in this chapter you'll use more often, and which techniques you'll use less often. Look back at your strategy to see which techniques help solve your problems. However, in continuous recruiting mode, you'll want to use the majority of this chapter's techniques, because they set up a continuous draw in the professional community.

Internal recruiters

Internal recruiters—those people within a company who focus specifically on recruiting new hires and on relocating employees already on staff—generally do not perform any HR functions other than recruiting. On occasion, I've had fairly good results with internal recruiters who have reviewed phone-screen scripts and helped me to choose appropriate sourcing strategies, making the interviewing experience both productive and positive.

Unfortunately, I've rarely had good results using non-specialized HR staff; more often, I find that internal recruiters actually derail me, pre-

---

[1] Robert Wendover addresses the topic of re-hiring former employees in his book *Smart Hiring: The Complete Guide to Finding and Hiring the Best Employees*, 2nd ed. (Naperville, Ill.: Sourcebooks, 1998), pp. 222ff.

[2] Recruitment strategies are treated extensively in Ed Michaels, Helen Handfield-Jones, and Beth Axelrod, *The War for Talent* (Boston: Harvard Business School Press, 2001).

venting me from finding the best-suited candidates, because they don't have the external contacts to source properly, and they don't emphasize what is specific to my open positions. Too often, internal recruiters look for tool and technology expertise or for advanced academic degrees, rather than for functional skill or for product-domain experience. If the internal recruiter is not sufficiently knowledgeable about your open positions, you do have some choices: You can try to educate the recruiter, you can make your tradeoffs more explicit, or you can filter résumés yourself without allowing the internal recruiter access.

When I do use an internal recruiter, I treat him or her as I would an external recruiter, and review all résumés myself, do my own phone-screens, and use the internal recruiter primarily to set up interviews.

Use money, not time, to attract suitable candidates.

A well-rounded sourcing strategy uses most of the above strategies in some way. If you have a substantial budget but not a lot of time, consider using a combination of the more costly sourcing techniques—such as print and media ads, external contingency recruiters, external retained-search recruiters, headhunters, and numerous nontraditional approaches—along with the time-intensive techniques. The following paragraphs describe these methods.

Newspaper ads

Many candidates prefer to begin their job search by reading Help Wanted ads in the Sunday newspaper. They like this time-tested approach because it gives them a way to review advertisements, do research on the companies, and then narrow their search to those companies they consider to be prime. When you use Help Wanted ads, make sure the ad copy will appeal to the best-suited candidates. Pay to have the ad set in a type size that is large enough for people to read and pick a font that is legible. If candidates can't read your ad easily, or aren't attracted to it in the first place, the best candidate could miss knowing about your open position.

Radio ads

Radio spots don't cost as much as newspaper ads, making it possible that advertising over the radio may be affordable for you. Make sure, however, that the radio station's demographics fit your hiring profile.

Internet ads

Not only can you place a job announcement on a Web-based résumé board, you also may be able to advertise there. You'll have to negotiate a recruiting budget with your management to determine whether advertising on the Web makes sense for you, but it is important to map your ad strategy to include investigation of this forum. In addition, if you're thinking about using Web-based advertising, consider placing ads on Websites that have demonstrated their effectiveness in reaching a specific geographic area, so that you draw candidates from the corresponding geographic area. Because many newspapers around the world work in tandem with Internet sites, your job can be advertised simultaneously in print and on-line, and you can choose how and where to source.

External recruiters

External recruiters can be expensive. They may require a fixed payment or a commission for each candidate you hire. Their compensation may be calculated according to a percentage or sliding scale, based on the candidate's annual salary. Whatever their requirement, their payment should be contingent on your hiring a candidate they bring to you: If you don't hire a candidate they propose, you don't pay them. It's that simple. Such external recruiters are sometimes called "contingency" recruiters because their payment is contingent upon the hiring of a candidate they have produced.

If you're going to work with external recruiters, spend time finding ones with whom you enjoy working. Invite them in to meet with you; show them around your offices so that they see the physical environment; explain what is different, or special, or strange, about your company and its culture and identify characteristics typical to the kinds of people you require.

External recruiters can provide feedback about your job description, especially when they compare your required skills, preferences, and qualities with the money you're willing to pay for a candidate. External recruiters also can tell you when you're offering too much—or too little—money for a particular position. They can warn you if the salary you're willing to pay is lower than salaries they've seen offered for comparable jobs.

If you're not sure whether to use an external recruiter, here are some guidelines to tell you when you could make use of such a person:

1. If you believe the best candidates are not currently looking for jobs, use an external recruiter. External recruiters know where and how to find the most suitable candidates, whether those candidates are in the market for a new job or not.
2. If you have numerous openings and you're not sure whether you can source a sufficient number of résumés to yield enough people to meet your hiring schedule, use an external recruiter.
3. If you're looking for someone with a unique combination of qualities, preferences, and non-technical and technical skills, use an external recruiter.

I've had satisfactory results with external contingency recruiters when I've done the following: I give them a complete job description and explain how this position is different from other positions with which they may have helped me recently. If they send me candidates who aren't quite right, I explain why the match is not right, and look over the next batch of résumés with the recruiter at my side to see whether he or she understands what I am looking for. I ask the recruiter to check my elimination questions regarding salary, potential start date, and so on, so I know in advance that the candidate is worth my time to phone-screen.

The external recruiter works for *you,* not for the candidate. He or she won't charge candidates because your company will be paying the fee. Whenever you have a fee-based agreement, discuss *how you like to work* with the recruiter. Search for ethical and knowledgeable recruiters and use them. They will be worth your time and money when the circumstances are right.

Some external recruiters are called "retained search" recruiters, "executive" recruiters, or "headhunters." They source senior-level, management candidates. Sometimes, these recruiters also take on searches for senior-level technical staff. When working with these recruiters, make sure you've identified what's unique about the position to be filled. Interview the recruiter so that you know that he or she understands your required functional skills, your product domain, and what tool, technology, and industry experience is needed.

If you will need to sign a contract with a retained-search recruiter to hunt for a candidate, read the fine print in the contract. Know whether you will have to compensate the recruiter regardless of how you find the candidate. Also, know whether the recruiter requires an exclusive-agent arrangement and, very important, when the contract ends. Ask for industry-related references to ensure that such recruiters know your field well enough to search effectively for your candidate.

When I work with external recruiters, whether they are contingency or retained-search recruiters, I discuss how we're going to work together. I explain that I will perform the phone-screen, set up the interviews, and negotiate the salary with the candidate. I also explain how I will keep the recruiter in the loop. Recruiters can be helpful, as long as you have planned how you will work together. Explaining how I work—at the beginning of the partnership—helps both the recruiter and me perform our roles.

Other nontraditional ads

Other campaigns to publicize a job opening may include such nontraditional approaches as hiring small planes to tow banners over beaches in the summer, skywriting, roadside posters and billboards, humorous ads, and numerous other delightful and creative ploys. Some companies advertise in magazines that they believe their candidates are likely to read but that do not treat the subject matter associated with the advertising company's product domain.

If you have help available in the form of professional copywriters in an internal marketing department, then you might want to experiment to see whether general-audience advertising works for you. Typically, such advertising requires a long lead time and can be expensive, but if you have the resources, see what works best for you and your company.

No matter which sourcing techniques you prefer, make sure you use some combination of time and money, along with the above practices, to reduce the risk that you won't be able to find someone to hire when you are ready. Don't rely on just a few practices, even if you're using external recruiters. Create a diverse sourcing strategy so that all those great candidates can easily find you.[3]

Once you've selected your sourcing mechanisms, you'll know how to focus your ads, but at the very least, make sure your Website is updated with current open positions and their descriptions.

POINTS TO REMEMBER

- The more time and effort you put into defining your recruiting strategy, the more easily you'll find qualified, suitable candidates.

---

[3] Discussion of diverse sourcing strategies can be found in Michaels et al., op. cit., and in Lou Adler, *Hire with Your Head: Using Power Hiring to Build Great Companies,* 2nd ed. (Hoboken, N.J.: John Wiley & Sons, 2002), pp. 248ff.

- Use a diverse set of recruiting techniques.
- Make sure potential candidates can easily find open positions on your company's Website or job page.
- Use an employee referral policy that brings in referrals but doesn't distract employees from their technical roles.
- The more exposure your ad has, the more important it is to include details in it that allow potential applicants to determine for themselves whether responding to the ad would be a good use of their time. Don't hold back information that either would encourage the right candidate to contact you or would discourage the wrong candidate from applying.

5

Developing Ads for Open Positions

"Oh, I'll just let the HR people write the ad. They'll know what to say."
—Hiring novice

Job descriptions help you screen candidates by making it easier for you to differentiate between the essential and desirable requirements for a particular job. The proper ad can further aid the screening process by offering a candidate an opportunity for a career, not just a job.[1] A well-written ad helps the candidate determine whether or not your open position is worth his or her response. Develop an ad that helps the most appropriate candidates respond to you.

You'll have to tailor your ad to your recruiting medium (electronic, print, or any of the other mediums suggested in Chapter 4, "Sourcing Candidates"). Use a job ad template to organize the information you'll want to include in the ad. Begin preparing to write the ad by noting the main attraction—that is, jot down the selling point of the specific job. Then, list the opportunity to be described by the ad—the activities the candidate will perform—and the essential qualities, preferences, and non-technical and technical skills, those details that should have been specified in the job analysis and job description, ready for you to pull into the ad. Conclude with contact information.

By presenting specifics in the ad in this way, you will help candidates screen themselves. That is, if they want a job with the selling point you've listed, and they are interested in what is shown in the ad as the main attractor, then they'll keep reading. If they read on and determine that

---

[1] The importance of distinguishing in an ad between a job opening and a career opportunity is treated especially cogently by Lou Adler, *Hire with Your Head: Using Power Hiring to Build Great Companies,* 2nd ed. (Hoboken, N.J.: John Wiley & Sons, 2002), pp. 236ff.

they possess the qualities, preferences, and skills described in the ad, and they want to perform the work, they'll send their résumé in response to the ad. Despite the screening your ad copy will provide, you'll probably still receive résumés that don't fit the open position, but you'll receive fewer of them with a well-thought-out, well-written ad than you would otherwise have received. The only time truly unqualified candidates don't screen themselves out is during a down economy, when just about anyone who can draw a breath and is looking for a job will send you a résumé.

Use a simple job advertisement template.

For written ads, I again use a simple template, as follows:

Job Advertisement Template

[Company name] is looking for a [job title].

Main attractor:

Deliverables and activities:

Essential qualities, preferences, and skills:

Contact information:

Main attractor: Define what will attract the kinds of people you want to hire. Consider corporate cultural-fit factors, and product and technology matches. Make the main attractor upbeat to draw candidates into the job. Every job has something that will attract potential candidates. It could be the job itself, the company, the technology, the people already in the group, or something else. Position what you think is most attractive to candidates at the start of the ad. When the company is the main attractor, start with facts about the company as noted in your job analysis:

Dynamic start-up company seeks a senior software developer . . .

Established, profitable company seeks a talented support engineer . . .

If the job itself is the main attractor, you could start the ad with a question that implies a benefit. Be careful to avoid questions that candidates are likely to answer No to, however; they won't keep reading if they do not like the implied benefit. Examples of possible opening questions follow:

Tired of using the same old approach to testing? Create a test environment you can be proud of . . .

Ready to take on a bigger challenge than designing Web pages? Learn to craft and build dynamic new products . . .

With technology as the main attractor, you might start an ad with a reference to, say, wireless computing:

Ready for the challenge of supporting wireless computing?

If the main attractor is the wonderful group of people already on staff, start with that fact as the main attractor:

Come join an award-winning team of technical writers . . .

Make the main attractor specific. Simply naming generally available, subjective benefits will not be particularly effective. Try to identify what is special about your job opening rather than just heralding a "great company," "strong technology," or "terrific people." None of these three "benefits" will be particularly effective in grabbing the attention of a potential candidate. Make your main attractor specific and compelling so that it forces the candidate to notice your ad.

If you have more than one main attractor, you can create multiple ads, and see which ads bring in the best batch of résumés. You also can use more than one of the attractors in one ad. You even can put *all* the attractors into a single ad. When I have multiple main attractors, I usually write and post multiple ads, and see which ad brings me the greatest number of qualified candidates.

It's worth spending time thinking about and describing your main attractor. The main attractor describes the *opportunity* for candidates—the "thing"—tangible or intangible—that differentiates your job from everyone else's.

Once you know what you want to start with, you can write the body of the ad.

Deliverables and activities: Use specifics from the job description template presented in Chapter 3, "Writing a Job Description," to build the middle of the ad. Describe the kind of work you want this person to be able to perform. Detail makes your environment real to the candidate. For example, you might write something like the following:

As the business systems architect, you'll lead the work-flow definition and product development of the new logistics system.

As a tester, you'll help find inconsistencies in our product line.

As a project manager, you'll manage all subprojects and report on project state to senior management.

When candidates read ads, they should be able to visualize themselves in the position you're describing in the ad. The more effectively you describe the qualities, preferences, and skills you're looking for, the more accurately the right candidate can identify with the position. If a candidate is a good match, he or she will be sold on the job before you even read a résumé.

Essential qualities, preferences, and skills: After the activities and deliverables, add the essential cultural qualities, preferences, and skills, possibly to include details about how the candidate will perform the activities and deliverables. Although you might want to hire a project manager with strong negotiation skills, it's not very compelling to write "Must have significant negotiation skills." Instead, try, "You'll negotiate with suppliers and with the project team about milestones and tradeoffs." Here are some ideas for making this part of the job real to candidates:

As the business systems architect, you'll lead the work-flow definition and product development of the new logistics system. You'll bring diverse groups of users together and facilitate the definition activity for the next release.

As a tester, you'll help find inconsistencies in our product line. You'll be working hand-in-hand with the developers, reporting problems as you detect them.

As a project manager, you'll manage all subprojects and report on project state to senior management. You'll organize the activities for this release and help plan the next one.

If you only mention the required number of years of tool or technology experience that applicants must have, you've given them no way to evaluate the merits of your position relative to other jobs. Although you'll receive plenty of résumés, most of the people won't fit, either culturally or technically.

Contact information: Complete the ad with specifics about how you can be contacted. I always include my e-mail address and fax numbers, but I intentionally omit my telephone number, and I make it explicit in the ad that applicants are not to contact me by telephone. Here's possible wording for this fourth part of the ad:

> *Please e-mail your résumé to hiringmanager@yourcompany.com, or fax your résumé to 999-555-YYYY. Refer to position PM103. No calls please.*

Always let candidates know how you prefer they submit their résumé. If you don't want to receive résumés by postal service or internal routing, say so. If you don't want recruiters to send you candidates, include phrasing such as, "Principals only" or "No recruiters." Specify the requisition number or the name of the contact person because that detail will help you determine what ad a person is responding to and also will help you determine which ads work best.

Write different types of ads.

In Chapter 4, "Sourcing Candidates," I described using multiple sourcing techniques and media to attract candidates. Different media can require different kinds of ads to be effective.

If you need to write multiple ads, follow the job advertisement template as you prepare each and every ad. Some possible types of ads and media are reviewed in the following paragraphs.

Ads for your company's Website

You have the luxury of *space* on a Website, not only for words, but also for graphics and pictures. There are several points to take into consideration when you post an ad on your company's Website:

1. *Make sure your Website is structured so that people can find your ad.* Too many corporate Websites make it hard for potential candidates to find listings of open positions. Typical problems are job pages that only a telepath can find; job titles that are meaningless to the outside world, such as listings that call for a business analyst when what is needed is a developer; listings that do not state a specific geographic location;

and Websites that cannot be accessed by certain browsers. Work with your company's Human Resources staff or with your Web developers to make the site work for you. Refer back to Chapter 4, "Sourcing Candidates," for additional Web-related information.

2. *Identify the main attractor for the particular audience.* The main attractor for a Website posting is not necessarily the same one that you'd use for a print ad. If people are reading the ad on your company's Website, they presumably are already interested in the company and are looking for ways they can join it; the main attractor does not need to repeat what they already know. Think about the audience and target it for this version of the ad.

3. *Make the important points easily visible.* Remember that you're writing for the Web, and that people won't be willing to invest much time looking for the information.[2] Although you have more space than you would for a print ad, don't make people work too hard to understand your ad by putting in irrelevant detail or by burying the meat under too much verbiage. Once the major points have been highlighted, feel free to add more information.

Websites are made more attractive and more dynamic if you can post photographs or renderings of your products, your workplace, a recent group outing, or some other picture or graphic that will help attract the most appropriate candidates.

Ads for a recruiting Website

There are numerous Web-based recruiting sites, some of which were mentioned in Chapter 4, "Sourcing Candidates." If using a Web-based recruiting site appeals to you, consider the following ideas when creating your ads:

1. *Create ads by following the instructions given at the recruiter's Website under the heading "For employers,"* or the like. Use the tips, templates, and keywords provided at such sites, which often offer kits to help you create an effective ad.

---

[2] Writing for the Web is given comprehensive treatment in Gerry McGovern, Rob Norton, and Catherine O'Dowd, *The Web Content Style Guide: An Essential Reference for Online Writers, Editors, and Managers* (London: Pearson Education, 2002).

2. *Read other companies' ad postings on recruiting sites to see how they describe and promote open positions similar to yours.* Think about how you will differentiate your ad from those ads. If the recruiting site e-mails position postings to candidates, make sure your subject line will stand out.

3. *Tailor your ad to the specific kind of candidate who is likely to go to each recruiting site.* Because each site will attract different kinds of candidates, you will want to consider who those candidates are likely to be. Candidates may use a general job site, such as Monster.com, to look for specific companies by name, searching, say, for IBM or Dorset House Publishing; or candidates may use a technical site, such as Dice.com, to look for a technically challenging job in, say, the aerospace industry. Use different ads for the different Websites.

E-mail ads and newsgroup ads

If your company uses its Website to gather customers' e-mail addresses, you may be permitted to access it in order to e-mail a targeted list of people, many of whom could be interested in seeing your job postings. In addition, you may be able to use the e-mail lists of such sources as local or national professional societies or technical newsgroups, if they will allow it. Look at the charter of the professional society, ask the mailing list owner, or read the newsgroup's frequently asked questions (FAQs) section to determine the rules and regulations pertaining to job postings.

Whether you wish to send ads to an e-mail list or post them on your Website, there are techniques to keep in mind:

1. *Use a subject line that means something,* such as "Job Opening: Tester Wanted."

2. *Consider sending just a brief teaser containing a URL instead of the full ad:* "Searching for developers who want to develop high-performance, low-defect telephony software. Access our URL for more information."

3. *Choose the main attractor for the particular audience.* People reading e-mail from professional societies may be interested in how your job will enable them to advance in their careers.

4. *Keep the ad to one screen-length so that people can read it quickly.* You don't have to be as brief as you might want to be for a

newspaper ad, but you still need to be brief enough so that you don't bore or lose the attention of potential candidates before they decide whether they want to respond to you.

5. *Use plain text for the e-mail message.* You don't need fourteen colors or twenty fonts and fifty exclamation points to tell people you're recruiting for a technical position. Sending highly formatted e-mail is likely to alienate the very people you want to recruit. If you have a visually pleasing ad, send the URL and leave the ad posted on a Web page.

6. *Avoid using attachments.* If you're sending e-mail to a sizable mailing list and don't know what variety of browsers recipients will be using, put the entire ad into the body of your e-mail message. To make it easier for yourself, you can draft the ad on a word processor and then transfer the text to a plain-text e-mail message. Remember that not everyone can or will receive, read, or open an e-mail attachment.

7. *Include a way for people to remove themselves from your list,* even if you're using a list for which people opted in.

Flyers

One-page flyers are a practical way to broadcast a job opening. You can post them on job boards, hand them out at professional meetings or conferences, or leave them in the lounge of the career center of your local college or university. In determining what to include on the flyer, first decide whether you will use it to advertise a single opening or multiple jobs. If your company has multiple open positions, you may want to create a flyer that features one job but lists the other positions. The more positions you mention, the greater your chance of interesting qualified candidates.

Generic ads

Generic ads describe the company, the technology, or the culture. They encourage people with general skill-sets (such as software developers) either to contact you or to visit your Website for more information. If your company plans to hire fifty or more people in a year, you'll need generic ads for continuous recruiting. If you're a technical hiring manager or other technical person, don't write a generic ad; ask your HR or marketing group to write that ad. It's probably not worth your time to perform generic hiring work.

Radio and television ads

Radio—and even television—ads are especially effective when you have numerous, similar openings; when you have multiple openings with the same job title; and when the economy is so strong that you encounter formidable competition in trying to attract the most qualified candidates. Such ads need to communicate a dynamic aura in a few carefully chosen words, and there is an art to writing them. When you write an advertisement that will be spoken on television or the radio or read aloud at a meeting by someone other than you, you will need to include more carefully selected information than would be required in a brief, classified ad. Following are three examples of when and how to prepare such ads:

- If you have a number of similar open positions that you need to fill, radio ads that target a particular demographic group can be helpful, as can television ads if your budget is sufficient or you have access to a low-cost, cable channel on which to advertise. For example, if finding entry-level people is difficult, you could develop targeted ads for a fresh crop of engineering graduates and place them on a talk-radio station or a sports channel.

- When the economy is especially hot and even when your industry itself is experiencing a boom-time during an otherwise depressed economy, you may have trouble filling open positions. Don't get discouraged, however, as a hot economy provides a great time to use a specific radio ad ("If you know how to test watchdog timers, we want you") in addition to the standard, generic ads your company may be in the habit of placing. If you're using radio in addition to other media to hire against a specific open requisition, then write the ad as if you're writing for the newspaper, but keep it to whatever number of words can be articulated clearly during the time allotted for the radio or television spot you have selected.

- When you're hiring many people with one skill-set or one essential job-performance qualification, your ads should target that group of people. For example, if you need developers or testers but you don't have any specific preferences about personality or working habits, you can include details about job deliverables and perhaps also add something about the company, such as its being a start-up or a leader in the field (see Chapter 2, "Analyzing the Job").

To see what can be done in such a situation, take a look at two drafts written for a thirty-second ad as part of a campaign to hire a large group of software developers. The first draft is close to one minute long, too long for radio, and contains obviously extraneous verbiage—but it is just a draft intended to get ideas down.

Ad Draft 1:

Looking for a project-oriented culture in a small company that is experiencing significant revenue growth? Join Walker Software, Inc., a supplier to the telephony industry, as a Software Developer. You'll work with other developers to design and enhance Walker Software's product line. Because we care about the performance, reliability, and quality of our products, we use simulations and modeling to predict product performance. In addition, we use numerous peer-review techniques to reduce defects, including design reviews, code reviews, and smoke tests. If you're interested in building quality software quickly, and you have at least four years of experience, including UNIX and C++, call us at 999-555-XXXX, or e-mail your résumé to hiring@walkersoftware.com.

The second draft takes about thirty seconds to read and is about right for a short radio or television script.

Ad Draft 2:

Looking for a project-oriented culture in a small company experiencing significant revenue growth? Join Walker Software—a supplier to the telephony industry—as a Software Developer. You'll work with other developers to design and enhance Walker's product line. If you're interested in building quality software quickly, and you have at least four years' experience, including UNIX and C++, call us at 999-555-XXXX, or e-mail your résumé to hiring@walkersoftware.com.

Radio and television ads may need to be even shorter than newspaper ads to meet time constraints. Ask an advertising sales representative or copywriter employed by the station at which you will place your ad to help you tailor it for the specific audience demographic.

Movie ads

Movie ads offer another good opportunity for exposure. Without too much difficulty, you can redesign generic, company-oriented ads to be projected onto the screen in movie theaters. Because time slots for such ads may give you more time than a print or spoken ad, and because you can incorporate images as well as sound, you usually can list specific titles for the openings you're trying to fill. Time may permit the ad to include a full description of the company, its culture, product line, and the like. Make sure that your ad runs as a preview for the kinds of movies your technical candidates are likely to see.

Newspaper ads

Newspaper ads can be challenging ads to write because they should be both attention-getters and concise. If you will pay for the ad by the word, by the line, or by the inch, organize and write the ad to maximize the biggest bang for the buck. A well-written and well-structured newspaper ad can yield the highest number of qualified candidates in the shortest amount of time, giving you the best return. A poorly conceived and sloppily written ad will provide only a costly lesson.

What you should include in a newspaper ad should be as specific as you can make it and still meet space and budgetary constraints. Imagine that you need to write an ad geared toward hiring a technical support representative for a real-time instrumentation company that is number two in its field, is profitable, and whose product has both hardware and software components. In addition, you know you need a person who will be able to deal with customers ranging from technicians in laboratories and factories to research scientists, and who also will be able to interact with hardware and software engineers and with the technical writers who will document how the product works.

No problem! Start with the main attractor, that key piece of information that will catch the attention of precisely the candidate you need. In this case, you think the main attractor is the hardware and software product combination in conjunction with the specific industry—real-time instrumentation.

Here's one way to begin the ad:

Tech. Support Rep

Want to discover and solve challenging problems at the junction of hardware and software? Real-Time Instrumentation is looking for people who want to use their investigative and diagnostic skills to discover problems anywhere in the product and eliminate them.

These first two sentences should attract people who know they must be able to solve problems, people who also aren't afraid of hardware or software. You'll need more information in the ad, however, than this main attractor if you are going to find the perfect candidate. You'll want information that will cause job-seekers to consider their own qualities, preferences, and skills, so they can screen themselves. One category you'll want to define may be the level of experience a candidate has.

If you're looking for people *with experience,* the rest of the ad might run as follows:

> *If you've supported real-time embedded systems and can show us how you've found and solved customer problems, or if you've developed or tested instrumentation products and you're interested in a different challenge, we want to talk to you. We're looking for people who can persevere through difficult problems, who want to delight our customers, and who can help move us into being the number one company in our field.*

If you're ready to try people *with little experience,* consider this ad copy:

> *If you can show us how you've persevered in the face of difficult problems, we want to talk to you. We don't care if you don't have technical support or instrumentation experience; we'll train you. We want to see how you solve problems and how you deal with customers who range from technicians to Ph.D. chemists.*

In the first example, you're looking for experienced technical support representatives. In the second, you're looking for smart people who can solve problems and be helpful. You'll attract different candidates, depending on the ad you place. In either case, end your ad with contact information:

> *Please send a résumé to hiringmanager@real-time.com, or fax it to 999-555-YYYY, attention Tech. Support Rep 101.*

If you can place only one ad but you need both experienced and inexperienced candidates, you might combine and consolidate wording as follows:

Tech. Support Rep

> *Ready for Tech. Support challenges? Do you want to discover and solve challenging problems at the junction of hardware and software? Real-Time Instrumentation is looking for junior and senior tech. support reps*

who want to use their investigative and diagnostic skills to discover problems anywhere in the product and eliminate them.

If you enjoy investigating problems and solving them, tell us how you've persevered through difficult problems. If you're an experienced tech. support rep, tell us what you've done. We don't care if you don't have technical support or instrumentation experience; we'll train you. If you like solving problems and enjoy dealing with customers who range from technicians to Ph.D. chemists, please send a résumé to hiringmanager@real-time.com, or fax it to 999-555-YYYY, attention Tech. Support Rep 101.

Shared space and multi-purpose ads

If your company is in the midst of hiring numerous people, you may have to share ad space with other hiring managers, or you may not be allowed to write the ad copy at all. If you must share ad space with other hiring managers, ask whoever is coordinating the ad to provide a sample of its layout. Also ask to see the non-job-specific copy. In most cases, if the generalized portion of the ad already states the company's main attractors, you can start your portion of the ad with position specifics.

If your company doesn't permit hiring managers to write their own ad copy when their ads are running with ads for other open positions, provide job-specific input to the copywriters in advance and ask to review the copy before it is posted, to help copywriters feature details you want included.

One-position ads

Individual ads, crafted to attract applicants to one position, work well for classified ads in the newspaper, or in some other written medium where brevity is required. If you are using alternative recruiting techniques in addition to newspapers, such as radio, or you're sending e-mail to a mailing list, you can choose different information, or include more information.

No matter which techniques you use for writing an ad, make sure you've started with what you think will attract the best-suited candidates. Ads must be designed to attract candidates, and you want to make sure you've attracted the right people. Adapt your ads to your recruiting techniques.

Case Study: Walker Software

Now let's see how Walker Software might handle a multi-purpose or shared-space situation. Omar, in Human Resources, has started to write ad copy.

Fiscally sound, prudent, software company has technical positions available.

Vijay, Walker's Director, and his managers took one look at that opening and stated that it was boring and would attract only risk-averse people. To Vijay and his team, the ad copy sounded tired and unexciting, except possibly to people who'd been laid off. Vijay and his managers don't want to attract candidates solely because they've been laid off and are unemployed. They want to lure dynamic, gainfully employed people away from their current jobs although they also want to attract out-of-work candidates if they are fully qualified for the job. The bottom line is, they want people who are excited about their jobs, people who want to contribute to the success of projects at Walker Software. They proposed the following alternative:

Looking for a small-company, project-oriented culture that expects significant revenue growth projected over the next year? Join Walker Software, a fiscally responsible supplier to the telephony industry, in one of these positions:

All of the managers can agree on this opening. After they explained their thinking to the HR rep, the rep agreed to change the start of the ad and gave Vijay and his managers each permission to write their own ads.

Dirk, Walker's Development Manager, wrote the following ad for a software developer:

Software Developer needed for telephony product. Work with other designers to develop and enhance Walker Software's product line. We use numerous review techniques to reduce defects, including design reviews, code reviews,

> *and smoke tests. If you're interested in building quality software quickly, and you have at least four years of experience, including UNIX and C++, fax your résumé to 999-555-XXXX. Refer to position SD 102.*
>
> Note that Dirk's ad references the requisite skills, qualities, and preferences he wants his new Software Developer to have. As Dirk also is focused on finding someone who wants to detect defects early, he's emphasized that in his ad.

Develop techniques for eliminating writer's block.

Writing an ad is difficult. If you're having trouble writing a particular ad, consider these ideas: You don't have to start at the beginning of the advertisement. You can start anywhere. If you know the essential cultural qualities, preferences, and skills that you require, then start with those. If you know how you want to finish, put that part in next. You can always come back to the beginning of the ad.

Don't be afraid to ask for help with the writing. After all, you're a technical manager, not a public relations or marketing specialist. If you have access to public relations or marketing people, they may be able to help you write the ad. Since they know what attracts your customers, they may be able to help you phrase the ad to attract suitable candidates.

If you don't have access to marketing people or technical writers or editors, try writing the ad using input from your hiring team. You'll probably come up with an ad you'll be happy to run.

If you've tried all these techniques and nothing is working, you can start the ad by noting, "We're technical people. We don't know how to write an ad, but we have openings, and we'd like to fill them." Your honesty will be refreshing and will attract candidates.

Make the ad memorable by offering a challenge.

Nothing is worse than a boring ad. Candidates will perceive the job to be dull if the ad is dull. Make your ads interesting by showing people a little slice of the job.

Great ads showcase the work people will perform, not just their technical skills. Show through your wording in the ad that you know that people are more than just a tally of their skills. The more you describe necessary technical skills in the ad, the less likely it is that a highly qualified candidate will think the job interesting. The more you include activities and deliverables in the ad, the more likely it is that you will receive résumés from strong candidates.[3]

Work with HR staff members when they write ads.

Human Resources staff members are there to help you. Your goals are the same: to hire the best-suited person or people for the job. If you work for a large company, your HR Department probably has professional recruiters who can help you write ads. If they can help, wonderful.

Most of the time, however, since neither you nor your HR rep is a public relations or marketing specialist, decide the best way to work with HR, and do the rest yourself. Or, hire someone who *is* a copywriter to write the ad.

Make sure outsiders review the ad.

No matter who writes the job advertisement, make sure a few people outside the writing team review the ad. You can ask for input from people who currently perform the job, or from would-be customers. Unless you're conducting a retained search in which recruiters will develop the ad and screen candidates for you, *your ad* will provide potential candidates with the first information they'll see about your position. Use your ad to make a great first impression.

Even if someone else in the organization writes the ad, you'll want to review it, to verify it's the kind of ad you want—that is, an ad that will attract the best-suited candidates.

Deliver the ad in person.

If you're personally recruiting candidates by attending professional meetings or other events that potential candidates might attend, you may have an opportunity to "speak" the ad. Don't read the ad aloud. Instead, develop a fifteen-second pitch to interest candidates enough to speak to you after the session or to encourage them to take a copy of the print ad.

---

[3] For more on these hiring strategies, see Adler, op. cit., pp. 7ff.

This speech can include the same wording you might use if you were sending e-mail and then pointing people to a URL.

If Dirk, our friend from Walker Software, had the opportunity to speak at a local-chapter meeting of a professional society, he could say, "I'm Dirk Jones from Walker Software, the soon-to-be-famous telephony company. We're looking for a developer who wants to develop high-performance, low-defect software. If you're interested, come see me later."

Once you've made the in-person connection, you have numerous options as to how to proceed: You could start an in-person screen, hand out copies of the job advertisement, point people to your company's Web page, and so on.

POINTS TO REMEMBER

- Start with the main attractor, the key piece of information that will catch the attention of candidates.
- Advertisements, like most other forms of writing, may require you to draft several versions and then submit your best effort for review.
- When you write ads, keep your readers or listeners in mind, and make your message easy to read or hear.
- Different media require different writing techniques to be effective.

6

Reviewing Résumés

Stanley: "Hey, Joe! Take a look at this résumé. Think we can bring this person in?"

Joe: "Well, maybe, but can you figure out what 'Led the widget-builder effort' means? Does it mean she's a technical leader, a project manager, an architect, or maybe an analyst?"

Stanley: "I can't tell."

Reviewing a résumé gives you your first opportunity to see whether a candidate's qualities, preferences, and skills match your job requirements and culture, but it does more than just that. It also gives you the chance to form a preliminary opinion about a candidate as an individual, not just as a set of attributes. Learning to review résumés *quickly* does even more— even though each résumé is unique, speedy review gives you an advantage over your competition because it helps you to detect the best-suited candidates in the shortest amount of time. Building a fine-tuned, résumé-review process that functions as a screening component can help you hire better and faster. I call this review process the *résumé filter,* by which I do not mean that this process filters résumés but rather that my review of the résumé itself enables me to filter the suitable candidates out from the unqualified candidates.

Correlate your résumé filter with the openings you have to fill.

Depending on how many people you're trying to hire and at what professional level, your résumé filter will need to be tightened or loosened to suit the occasion. The tighter the filter, the fewer résumés will make it to your

103

phone-screen. Similarly, the tighter the filter, the more useful it is for hiring senior-level staff members, a very small number of people, or a very specific set of skills. Carrying this concept still further, the tighter the filter, the more résumés you'll need to read, and the longer it will take you to find someone. Conversely, the looser the filter, the more candidates you'll phone-screen. So, there clearly are advantages and disadvantages to each adjustment of your résumé filter, but it makes sense to match the filter to the job and to your time frame (that is, adjust the filter to complement the amount of time you have to hire someone).

Because of the type of hiring I do, I've used a tight filter only once in more than fifteen years of interviewing. On that occasion, I was looking for someone with specific industry experience in health care, exceptionally strong technical skills, and some functional management skills. The combination of technical and management skills, as well as the need for domain expertise in health care necessitated a tight filter. In most situations, I tend to use a loose filter so that I can evaluate whether a candidate who has most, but not all, skills, or who has a close, but not perfect, cultural match could fit the position. A filter that is too tight may eliminate some great candidates.

No matter how tight or loose you make your résumé filter, as soon as you finish reading a résumé, evaluate that candidate's suitability in terms of a "Yes, he's a strong contender," "Maybe okay, but not really close to the ideal," or "No match at all." Sort the résumé into the applicable Yes, Maybe, or No pile. This sorting allows you to scan the résumés quickly, pursue the best-matched Yes-pile candidates, come back to re-read résumés for the borderline Maybe-pile candidates, and eliminate the unqualified No-pile candidates.

I use the Yes, Maybe, and No categories because doing so allows me to easily distinguish the "most likely" candidates from the "absolutely not" candidates. After you've scanned a dozen résumés of each kind, you'll quickly be able to differentiate the Yes-pile candidates from the No pile, but, if you're anything like me, you may read certain résumés and have trouble deciding whether the candidate is or is not suited to the job. Because I deal primarily with hiring technical people who may not know how to best present themselves in a résumé, I read résumés with an eye to giving each person the benefit of the doubt—which leads to the Maybe pile. I re-read the Maybe résumés after I've phone-screened the Yes candidates but still have not filled the opening, to see if I've missed a diamond in the rough.

Start reading each résumé at the top.

I read résumés by starting at the top, and work my way down to the final line. This sequence is not the only way to read a résumé—some hiring managers and recruiters go straight to the experience entry or to a candidate's academic history—but I believe that most candidates top-load their résumés and cover letters with what they think is the most relevant information. Seeing what a candidate thinks are his or her strong points can tell you a lot about the candidate. As unlikely as it may seem, some hiring managers read résumés starting at the end, but I discourage this practice because it can unfavorably color your impression of the candidate with outdated or irrelevant information, such as schools attended long ago or classes taken that no longer mesh with the candidate's career path. As I read, I make a mental note that I later will compare with the actual job's requirements in each of the following areas: work experience, position objective, strengths, education, and professional training. I use items in the résumé to *include* candidates, and only use items to exclude candidates if the items match my elimination criteria.

> *A True Story*
>
> Don, a hiring manager I briefly talked with at a professional-association meeting some years ago, told me that for many years he had taken pride in his ability to identify unsuitable candidates simply by looking at the last entries on their résumés. Don regularly rejected candidates out of hand based on what he saw at the end of a résumé, but would then read each rejected résumé from top to bottom to detect what was wrong with the already rejected candidate. He told me that he typically spent a month or more filling one junior- or mid-level position.
>
> Not surprisingly, Don's boss eventually lost patience with Don and fired him, explaining that he was too slow finding people to fill the open positions, and that she couldn't wait for him to find the perfect candidate.
>
> Don said that when he started his search for a new job, he discovered a company that seemed to be a perfect fit for his abilities. He spoke to the HR rep, who asked him to send his résumé to the hiring manager. Almost immediately, the

hiring manager rejected Don's application and telephoned him to let him know there'd be no interview. When Don asked why he was not viewed as a viable candidate, the hiring manager explained that one of the courses Don had listed on his résumé wasn't relevant to the job. Don protested that the course was just one of numerous courses he'd taken, and asked the hiring manager whether he had noted a particular piece of work experience Don thought the company would be interested in. The hiring manager replied, "Oh, no, that must have been at the beginning of your résumé; I only read from the end. I didn't even get that far after seeing that the course work was completely inappropriate for this job."

As we finished our conversation, Don told me he had learned an expensive lesson that day, and consequently had taught himself to read each résumé in full, always starting at the top.

Look for more than appears in print.

When I review résumés, in addition to starting at the top, I look at three areas: 1) technical experience, including functional skills, product-domain expertise, and tool skills; 2) industry experience; and 3) patterns of behavior. Items 1 and 2 are relatively easy to make sense of from a résumé, but item 3 is more difficult to discern. What I mean by trying to see behavior patterns is that I look for indicators that a candidate has weak areas that need explanation. Perhaps he or she has jumped from job to job in a short period of time, or has switched academic focus too often. Because I may not be able to determine fairly what the behavioral patterns in the résumé indicate, I make a point to ask questions to learn about the candidate's reasons behind the patterns.

Consider your fellow hiring managers' staffing needs while you review.

As you review résumés, remember that you're probably not hiring in a vacuum. Even if you're the only one with an open position, your fellow hiring managers may be using networking and continuous recruiting techniques so that they can hire people quickly when they have open requisitions. Or, if your company has numerous open positions, the company's other hiring managers also may be reviewing résumés. Bring possible candidates to their attention. I've found development candidates while recruiting for testers and support staff—and vice versa. When you give your peers candidate leads, you'll benefit because you will have given your peers reason to remember you while they're reviewing their own stack of submitted résumés or while they are out meeting people in the field or at conferences.

Read the cover letter or e-mail.

Read the cover letter or e-mail the candidate sends you. A good cover letter or introductory e-mail will not just mention your position, it also will relate the candidate's interests and experiences to your job. Do not be fooled, however, by a candidate whose cover letter relates his or her interest and experiences to a job or company that *only seems like yours,* as such letters indicate that the candidate has not done any homework about your particular company, but has sent you, and possibly dozens of others, a badly tailored version of a generic letter. It's up to you to decide whether a candidate who takes such shortcuts in preparing a job application and in researching your job openings may be worth further pursuit.

When I receive a résumé that is accompanied by a generic cover letter or a generic introductory e-mail but which neither explains why the candidate fits the job nor identifies which job the candidate wants to pursue, I wonder why the person bothered to write a cover message at all. In such instances, I ignore the cover message and focus on my review of the résumé—but I start with a slightly negative feeling about the candidate. When I receive applications through the mail or delivered by hand, I ignore the cosmetic aspects of the application—such as the weight or quality of the stock the cover letter or résumé is printed on, or whether the stationery is embossed or engraved, or whether the return address indicates a posh neighborhood or one from the wrong side of the tracks. I keep in mind that material aspects such as the color of the paper or its weight or where the candidate lives are never an indication of how well-suited a candidate is for the open position, and I focus on his or her qualifications, skills, and preferences.

Look for a work summary.

Many candidates state what type of work they think they are best at, or what work they want to perform, near the top of their résumé. Use the work statement to help yourself form an initial impression of whether the candidate's experience fits your needs.

If the candidate has provided such a work summary statement, I search for information on ways in which the candidate has helped a project, saved the company time or money, or provided some other significant and tangible benefit. The summary statement should match the experience. The following worksheet illustrates the way I record a candidate's statement of achievement and correlate it with his or her technical experience.

Summary Statement	Experience Focus
"Senior developer who is skilled at finding and fixing GUI problems, and who excels at determining how users use the system instead of designing what they say they want"	GUI design, requirements elicitation
"Excellent code base organization skills"	Configuration management
"Excellent ad hoc tester"	Functional testing skills

Worksheet 6-1: Work Summary and Experience Correlation.

If a summary looks like a rehashing of the candidate's work experience, I skim the summary and go on to read the candidate's statement of career goals and objectives.

However, sometimes, technical people don't write their résumés to support their summary. They may well have performed the work indicated in the summary. It is also possible that they may have misrepresented the work. Keep reading to make sure the résumé supports the summary. If I can't find any indication in the body of the résumé to support the claim that the person performed the work as reported in the summary, I place the résumé in the No pile and move on.

Compare the candidate's stated objective with the job description.

I review a candidate's stated objective to see whether he or she is looking for a position that is similar to or exactly matches the open position. When reading objectives, I compare what the résumé contains with what I have written as the description of the open position. I ignore generic objectives, such as "a challenging position in a dynamic organization." However, if the candidate uses a specific title, such as project manager, test lead, level-3 support engineer, or senior writer, I keep that objective in mind while I read the résumé. I check to see whether the candidate lists experience commensurate with the experience I expect of that level of work, while keeping in mind that the candidate's job titles may not match mine.

If the candidate's stated objective doesn't match your needs, don't automatically reject the résumé. For example, if a candidate for a management position has not yet been promoted to being a manager but is geared to make that move, and the candidate's work experience looks reasonable as preparation for a management position, I make a note to discuss the objective during the phone-screen and I put the résumé in the Yes pile. However, if a candidate is looking for a management job and has been a manager before, but I have only staff positions open, then I usually conclude that the candidate is not a match. I make an exception if the cover letter with the résumé indicates something like the following: "I originally was looking for a management position, but I saw your ad, did research on your company, and am excited about your technical lead position. Please consider me for this position."

While hiring for a technical lead position some years ago, I received a résumé from a first-line manager whose cover letter simply stated, "I'm looking for a position in which I can use more of my technical experience. Please consider me for this technical lead position." Although she had included little detail in the way of a work summary or statement of objectives, some of her work experience did match the opening, so I interviewed her—and I hired her. The position was a good match for her overall problem-solving and teamwork skills. She was happy and successful in the position and chose to stay a technical lead; to this day, she has not moved back into management.

As important as the match between objective and experience is, be sure to look for the special candidates who truly do want to change focus. It is a fact that many technical people want to take on positions of leadership and think that management positions are the only way to lead, but they don't really like being a manager. If you have a position that offers leadership and not management, you may receive résumés from managers who are unhappy with the management aspect of their job, and who may be ideal for the technical lead position.

Look for an objective, and see if that objective matches what you want in your position. Don't automatically reject candidates if they look too experienced for your position or have too much training or education. Your position may be right for them, and they may be right for you. Similarly, be open to candidates with excellent credentials and references even if they seem not to have quite the right experience or training.

Correlate the candidate's work experience with your open position.

When I review a résumé for work experience and accomplishments, I look for the context in which the candidate accomplished the work. I note how the candidate's work context is similar to or different from my open job's context. I start my comparison by noting similarities or differences in projects and companies.

If you take this approach, keep in mind that small projects tend to have less-formal communication channels than are common on large projects, and determine how heavily project size should be weighted. A candidate who has only performed technical work on a few six-person projects may not be the best choice to fill the project manager position on a fifty-person project—the context is too different. People who are accustomed to the accoutrements of large companies—such as administrative support, a training budget, and company-supported or company-supplied training— may have trouble making the transition to a start-up. The candidate may be capable of the transition, but if his or her work context has been substantially different from yours, you won't be able to tell this simply by reading about the work experience. If you place the résumé in either the Yes or Maybe pile, you can clarify this during the phone-screen.

A candidate's experience cannot be rated either good or bad; it is simply experience within a context. I try to understand the context of the candidate's work experience as I review the résumé, looking for qualities, preferences, and skills defined in the job description, comparing that context to my work context.

Evaluate tool and technical expertise when hiring technical staff.

When reviewing a résumé for technical expertise, I try to determine whether the candidate has enough relevant technical background to perform the job. If I have a list of specific requirements for technology, such as two years of C++ experience or five years of Java experience, I look for the reasons why those specific requirements are listed and then assess their importance for myself. In Chapter 2, "Analyzing the Job," I suggested that hiring managers specify why a certain number of years of particular experience is required. Along with that, a hiring manager must specify how he or she will detect whether a person has that experience. If you haven't determined what "two years of experience" means, define it now, so you can look for the specific experience you want to see.

I don't worry much about whether a candidate has language skills or other specific technical skills when I review résumés. Most technical people are able to pick up another language if they've already learned a couple of languages, especially if at least one of the known languages is functional and one is object-oriented. In my experience, technical people can relatively easily learn the technical skills necessary to use a specific operating system, the particulars of a database, or the technical environment.

If a candidate has not had experience with your specific technology, look at whatever formal education and experience is listed to see whether the candidate has had preparation that could allow him or her to succeed with your technology, especially if the technology is new.

For example, if your product is a real-time, embedded system written in an object-oriented language, you'd look for experience with real-time systems, or embedded systems, or object-oriented languages. A candidate who only has experience with batch-run, transaction-processing systems is not necessarily a good candidate for your position. However, if the candidate has learned several new technologies easily, the candidate may be adaptable enough to be successful in your organization. During the phone-screen and interview, ask the candidate about his or her ability to use previous knowledge to work with the new technology.

The key question for the hiring manager is whether the company can manage the start-up costs, including training, for a candidate whose technical skills are not a direct match. In my experience, candidates with a minimum of technical skills who fit the organizational culture are preferable to candidates with excellent technical skills who don't fit the organization.

When you screen résumés, make sure you're not screening based on just tools or technology experience. It's very tempting to use tools or technology as a résumé filter, but there are several reasons I've come across in recent hiring efforts not to do this:

1. *Tool experience is not counted the same way by everyone.* One candidate may not count her testing-tool experience because she's had less than a year of experience with any of the current testing tools. Another candidate may claim tool experience because he saw a tool demonstrated at a conference and used it once. A candidate may opportunistically use tool acronyms to force a résumé past the automated résumé screeners that select candidates based on their tool experience. Such candidates are not presenting you with an accurate perspective of their tool experience. Worse yet, you may miss the candidate with relevant, if limited, experience.

2. *Tool experience is not an indication of functional-skill or product-domain capability.* Not all tool experience equates directly as a fit with the rest of your technical requirements. If you screen solely based on tools, the candidate may not have the technical ability to do the work you require.

3. *Tool experience doesn't tell you anything about the personal qualities of the candidate.* As we worked through the job analysis in Chapter 2, "Analyzing the Job," we identified candidate qualities, preferences, and skills—all of which are independent of tool or technical expertise. If you screen only on tools, you may miss wonderful candidates—people who can learn your tools and technology easily, and who will fit easily into your culture.

4. *Tool experience is transferable to alternative tools.* Once a person has learned one tool in depth, it's comparatively easy for him or her to learn another tool. Tool knowledge is transferable across tools. If you screen résumés based on one tool, say, Microsoft Project, and the project manager has experience with, say, Primavera, you could miss a candidate with suitable experience.

5. *Tool experience is comparatively easy to acquire.* It's easy to learn how to use almost any tool at the superficial level, which is the only level you can assume from reading a résumé.

Look at the kinds of applications the candidate has worked on before. How similar is the application's context to your context? I'm reluctant to interview applicants for a position working with a real-time, embedded system if they've only worked on batch, financial, business applications, a middleware application, or just the GUI part of another application. If a candidate worked on real-time transaction processing, I would put the candidate's résumé into the Maybe pile for a real-time, embedded systems developer position, because the applications share some context. However, different product experience doesn't mean I automatically reject the candidate; it means I think about what other qualities the candidate can bring to the environment.

The products people work on influence how they think about their work, especially with regard to implicit product requirements. Someone who is used to thinking about high-performance, high-availability systems may have trouble dealing with an environment in which it's preferable to release a product on the publicized release date—despite the fact that the product has a significant number of known defects—than to delay its release. Candidates with a variety of product experience may be more adaptable than other candidates. Can you trade off adaptability for specific technical expertise? The answer will depend on your context: the job, the people already in the group, and the organization. Go back to your strategy and review your needs: Do you need someone who has the right qualities, preferences, and non-technical skills; or someone with the appropriate technical skills and the ability to learn quickly; or some combination? Your strategy will help you make tradeoffs.

Evaluate a management candidate's ratio of management-to-technical experience.

Managers of technical people don't have to be technical stars. Managers of technical people do need some combination of managerial and technical expertise, however. If you are searching for technical people who want to make the transition to a management position, look for evidence of managerial or technical leadership experience in the candidate's résumé. Use that evidence as the basis for asking phone-screen questions about the person's readiness to manage. If you're looking for a manager, and you have candidates who are not yet managers, decide whether you're willing to manage the risk of an untried manager in this position. See Chapter 14, "Hiring Technical Managers," for an additional discussion of the ratio of management-to-technical experience you should require.

Hiring a new manager from outside your organization is also risky, because that person may not have the industry expertise you require, and almost certainly will not have needed domain expertise. Hiring a new manager from within your organization is less risky, although he or she will still require coaching for the particular culture.

When I recruit managers, I'm careful not to make too high a level of technical expertise a requirement for the management role. Generally, I don't expect managers to have superior technical expertise; I expect them to understand each product's technical issues and to be able to hire qualified technical staff members.

Once you're a manager, it's difficult to enhance your technical skills, because you're doing management work. Great technical leaders generally have trouble being outstanding managers, because if they're doing technical work, they're not managing. Certainly, there are those candidates who are able to be both technical leaders and great managers, but those people are rare and, in most cases, expensive.

When recruiting a manager, make sure you've sufficiently analyzed the job so you know what problem you're trying to solve. Ask whether the major challenge is technical work or managerial work. Will the mix change over time? See Chapter 14, "Hiring Technical Managers," for more information.

Know the reasons behind multiple career or job changes.

When you filter résumés, note how many jobs a candidate has had, and at how many companies. Whenever a candidate has changed positions in his or her company, I make note of whether the move was lateral, up, or down. Whenever a candidate has worked at numerous companies, I try to detect whether the person's knowledge and value have increased with each change or whether he or she has continued to perform at the same level in the different jobs.

A candidate who has changed jobs frequently within a company or moved from company to company within an unusually short period of time may have great reasons for doing so. Such mobility can mean one or more of the following:

- The candidate is very capable, and seeks out challenging work.
- The candidate works for a company that has a development program in which employees are eligible to be rotated through a series of positions to give them wide exposure to different facets of the business, varied experience, and a broad network of peers.
- The candidate has chosen projects that complete quickly or are canceled before they finish.
- The candidate has worked for companies that experienced departmental or company-wide reorganizations, down-sizing, or wide-scale layoffs.
- The candidate is unmanageable, and each successive manager can't wait to move this person elsewhere.

Be wary, however, of making any assumption about why a candidate has listed frequent job changes on a résumé. People change jobs for various reasons, and it is a good strategy to ask the person for the reason. Generally, I ask for the reason during the phone-screen but sometimes I will wait until the face-to-face interview so that I can see how the person handles the question. As you gain more hiring experience, you will develop the ability to read a person's body language.

As you probably suspect from the preceding, I rarely rule someone out of a phone-screen just because he or she has changed jobs or companies frequently, but I also don't rule people out because they've stayed at one company for a long time. Their reasons should be discussed.

Two True Stories

> Some years ago, I phone-screened a developer whose résumé indicated he had worked for six companies in five years. As my first question, I asked him why he'd changed jobs so often. He was quick to answer: "For the first change, I was recruited out of my previous company, but then the new company lost the contract six months later, so I was laid off. During my outplacement, I was hooked up with another company, but it couldn't figure out how to ship the product, so it laid me off a year later. Then came the Internet boom—so my next two jobs lasted just eight months each as my new companies grew and shrank. I was desperate for a job I could get some traction on, so I thought the energy company was a good choice. Who knew it would get caught in the recession?" This poor guy made a number of bad choices, but he wasn't responsible for his companies' failures. He had thought out each move, but was a victim of bad timing and bad luck. When he came to me, he was desperate for a job he could remain in for years.
>
> At another point, I phone-screened a candidate who'd been at a large computer hardware company for more than twenty years before he was laid off. From his résumé, it looked as if he'd worked at similar jobs in different groups during the previous eight years. When I phone-screened him, I asked him what was similar about those jobs, and also what was different. Although I expected to hear a contrast of the roles between the jobs, the candidate replied, "Oh, there was nothing really different; it was the same job day in and day out. Nothing new." I asked whether there was something specifically challenging about any of the jobs, and he said, "No. I wanted to keep doing the same thing, so I did." He had managed to acquire the same year of experience eight times. I concluded that he had learned nothing new in the previous eight years, and wasn't a good fit with my high-energy, high-initiative culture.

If you only look at how long or short a time a candidate has been employed in a particular job, you may miss what is really important to know—the reason for the change. Use job change as a starting place to ask questions in a phone-screen.

Determine the reason behind an employment-history gap.

Candidates may have gaps in their employment history for many possible reasons, some of which are indicators of the candidate's ability to do the job and some of which are irrelevant. Others may be merely clerical. Some plausible explanations follow:

- The candidate wanted to experiment with a different kind of work that's not relevant to the current job search or career objective.
- The candidate served as a volunteer, working in the local school system or in a hospital for a period of time.
- The candidate was laid off and took time to find a new job.
- The candidate stayed home to care for children or an ailing spouse for a few years.
- The candidate mistyped the dates listed on the résumé.

When you see unexplained gaps in a candidate's employment history, continue reading the résumé. If the candidate looks like someone you'd like to phone-screen, then place the résumé in the Yes pile and plan to ask about the employment gap during the phone-screen. If the candidate's experience isn't right for the Yes pile, then reject the candidate without worrying further about the employment gap, but don't just reject a candidate because he or she has an employment gap. Anyone who's worked more than a few years probably has lived through some sort of job recession or personal time of crisis, and the gap in employment should not be too heavily weighted. The reasons are what matter.

Look for signs of merit-based promotions and initiative.

Some people have learned not just the strictly technical part of their job, but also how the products work. Such people may be domain experts for their product or experts in their industry. Others who list the same detail on their résumés may not have learned the same things, and, despite twenty years' working in the field, may have only five years of relevant experience. The point is, you must learn how to read between the lines of each and every résumé that crosses your desk.

I look for indications that the candidate himself or herself has taken the initiative to learn as much as possible about the job or the industry listed on the résumé, and is prepared to take on increasing levels of responsibility. A candidate whose company has sent him or her to numerous seminars, conferences, or even for post-graduate study may have the initiative you want—or he or she may only have done as instructed. Use the phone-screen to determine the reasons behind a candidate's advancement.

Look for indicators of cultural fit and of assumed responsibilities.

The culture in which the candidate is currently working or has worked will shape his or her expectations about how to work in your organization. I look for clues that help me assess cultural fit when I filter résumés. One piece of information I seek is how the person describes his or her work experience.

If I'm recruiting for an environment in which there is high interdependence and a well-oiled team, I look at how the candidate's current organization uses teams. I may phone-screen a candidate who doesn't describe teams on the résumé (especially if the résumé contains other context-appropriate experience), but I make a note to focus on the issue of teamwork during the phone-screen. Clearly, I will need to scrutinize the candidate's responses to questions that screen for how well the candidate has worked with others and how well the candidate gives and takes direction from his or her peers.

Some organizations use the word "team" to describe a group of people with the same function; others use "team" to describe a group of people working closely together. Key words such as "team member" or "with a team" help you determine whether someone has experienced a true team of people working closely together. I also look for phrases such as "contributed to" or "part of a team." I ask questions during the phone-screen to see what this candidate did and what the team did. My goal is to get a good idea of the person's experience, asking questions geared to reveal whether this person was a driving force on the team, whether the person took direction from other team members, or whether the experience was something in between. Don't be afraid to be direct in asking whether people on the team worked interdependently or independently.

Assess personal qualities and problem-solving skills.

When reviewing a résumé, I look for areas that I should probe in a phone-screen, to see what kinds of problems this person has solved and how. Did the candidate see a problem and fix it? If there is an indication that the candidate did just that, I might ask, "What were the consequences of fixing that problem?" It can be difficult to see areas to probe, but you can look for several clues in a résumé:

- For both junior- and senior-level staff members, I look to see whether the candidate made changes to his or her work processes or work environment. Look for phrases such as, "modified build process to make builds faster."
- For junior-level people, I look for indications of initiative, showing that they've tried to figure out how things operated when they first started working. Many times, junior-level staff members try to make things better for the next set of people. Look for phrases such as "created a list of known problems and their solutions for the tech. support knowledge base" or "documented build system" or "reorganized library for faster compilations and builds."
- For senior-level people, I look for examples of how they handled difficult projects, difficult customers, and difficult problems. Did the project manager manage the critical path, or did it manage him or her? Did the writer extract the technical issues from the developer or the requirements, or did he or she rely on the developers to explain the issues? Did the tester determine ways to test the product, even early in development? Look for phrases such as "managed the critical path" or "extracted architecture definition from developers," or for phrases that include other action verbs that describe the work a candidate performed on a project.

It's difficult to get a full picture of the candidate's work just from reading a résumé, but if you have additional questions after reading the résumé about what tasks the candidate performed at his or her work, then the résumé should be placed in either the Yes pile or the Maybe pile.

Assess education and technical skills in terms of the open job.

I first look at a candidate's educational history for two things: Does this person have the basic knowledge I'm looking for, and has he or she continued to pursue some form of learning? If the formal academic degree doesn't show that the person has the necessary background for the job, I look again at the candidate's work experience. Does it give me evidence that the person can understand the context of the position I need to fill? Is it worth conducting a phone-screen?

I'm usually more interested in discovering whether a candidate has taken job-appropriate workshops, professional seminars, or training courses, and has shown an interest in continuing his or her professional education, than I am in knowing whether he or she has earned a Bachelor's degree or higher. Of course, that way of thinking may not be appropriate to your situation, but I hope you will still take my point that academic degrees are but one measure of a person's competence.

As noted earlier in the book, I don't give much weight to professional certification, and I never list certification as a requirement for a position. The only assumption I believe I can safely make when I read that a candidate is professionally certified is that he or she had some specific book knowledge at the time of the certification. When I see two candidates with matching qualifications except one has achieved certification, I don't allow that fact to sway me in favor of phone-screening only the certified candidate. As always, I make the decision of whether to phone-screen a candidate based on his or her match with the job description, qualifications, and experience; once I have determined that the candidate is a viable applicant, I may ask him or her to talk about an applicable certification or about his or her education. I do this more to get a fuller sense of the candidate as a person rather than to weigh one person's accomplishment against another's.

Put typographical and other clerical errors in perspective.

When I review a résumé and see typographical errors or common mistakes in grammar or punctuation, I don't automatically throw the résumé into the No pile. How damning the error is depends on the kind of job the person wants. I generally ignore two kinds of error:

- *Commonly confused or misused words:* Résumés of candidates who list their job title as "Principle Engineer" instead of the correctly spelled "Principal Engineer" are not sent to the No pile. I recognize that people regularly misspell this title, and even though I'd like to cure the world of this fault, I'm not going to penalize a candidate for using what well may be his or her official company title on a résumé. In other cases, a candidate may cite his or her contribution to a book as a "Forward" when I know the correct word is "Foreword." Others confuse "affect" and "effect," "compose" and "comprise," and on and on. But unless it is my job to hire technical writers, in-house editors, or the like, I ignore these errors and focus on other qualifications.
- *Typos in résumés generated by external recruiters:* Many recruiters re-type a candidate's original résumé into a word processor in order to print it out on the recruiter's letterhead, but this practice can introduce typos not found in the orig-

inal. Typos also can creep into a résumé because recruiters sometimes dramatically change the contents, emphasizing what they think will sell the applicant—a practice that clearly misrepresents the candidate but that also can introduce typos. If a recruiter-generated résumé has typos, I assume the candidate would have known better.

I do *not* ignore typos of the following ilk:

- *Typos in résumés submitted by a writer or a tester:* Many writers have trained themselves to look for typos in their work, but don't always remember that a résumé is one form of writing. Typos in a writer's résumé do give me reason to pause, although I do forgive typos in e-mail because e-mail is an informal, rapid-fire way to communicate and I believe it is harder to carefully proof or spell-check e-mail than other types of written communication. So, it bothers me greatly when writers fail the test, but it also worries me when a tester misses typos—after all, a tester tests other people's product. If the tester doesn't test his or her own work product, what kind of work will that tester release?

A True Story

Because I tend to screen candidates *in* rather than out, I follow this accept-rather-than-reject approach even when typos are present in the résumé. This approach proved to be advantageous for one of my colleagues many years ago when reviewing a résumé submitted by a technical support candidate. Unbelievably, the candidate had misspelled the words "compiler" and "environment" on her résumé. The hiring manager suspected that typing proficiency was the culprit rather than ignorance, neglect, or incompetence, and interviewed the candidate. Based on her other qualifications and the interview, he decided to hire her, despite the typos. The decision proved to be the right one—she excelled at the company for more than five years, working first as a technical lead and then as a first-level manager in the tech. support and test groups.

If a résumé is *riddled* with typos, I probably won't phone-screen the person, because I want to find people who have both the ability and the will to review their work.

Evaluate résumé items in terms of local and national hiring laws.

Most people know that, in the United States, it's illegal to discriminate against candidates on the basis of marital status, race, religion, gender, health, age, and so on (check your own jurisdiction for laws pertaining to hiring practices). But not everyone knows that it's also illegal to discriminate *in favor of* candidates based on these characteristics or preferences. When candidates include details about those characteristics or preferences on a résumé, they are including information that you can't use in the employee selection process.

Wherever you're hiring people—whether inside the United States or outside—make sure you're fully aware of all gradations of local law. This knowledge will guide you as you determine what's appropriate to ignore, and what can be used when reviewing a résumé. Never assume that the practices you've always used will be universally appropriate. I learned this firsthand when I taught an interviewing workshop to an international class. Midway through the morning, a participant from Europe commented, "I start reviewing the résumé at the picture, and work on down." I was surprised by this and remarked that using a person's photograph to evaluate candidates is illegal in the United States. The gentleman responded that looking at a candidate's photograph as part of the screening process is encouraged in his country partly because it helps potential employers to see the candidate *as an individual* earlier in the screening process than they would otherwise. He added that since it is much harder to fire people in his country than in the United States, employers want to make sure they know as much about the person they are hiring as possible, and that includes knowing what the person looks like. So, although taking this approach would be ill-advised in the United States, the point to be made here is, Know and obey your local laws, practices, and customs.

My approach is to concern myself with whether the candidate can do the work, can come to work when needed, and can be productive. Once assured of those performance items, I can ignore personal information on a résumé. However, if you need to know something about the candidate *that is specific to the job you need performed*, and you can't find it on the résumé, ask for the information in the phone-screen and in the interview.

Evaluate each candidacy using your résumé-review process.

Review each résumé quickly and decisively. Although you may have figured this out from all the preceding sections, let me reiterate how I use the Yes, No, and Maybe résumé piles. If the candidate looks like a great match, the candidate goes into the Yes pile. If the candidate doesn't list the requisite skills on his or her résumé and I can see nothing that hints at the presence of those skills, I put the résumé in the No pile. If I have reservations about the candidate but feel more positive than negative about what I have read, I put the résumé in the Maybe pile.

Next, I phone-screen candidates in the Yes pile to winnow the selected set of candidates down to a few to interview in person. I notify the Maybe-pile candidates to let them know that they are being considered but that they do not have all of the required experience. I send a letter to the No-pile candidates to let them know they are not being considered further for the specific position (for more about how to do this without trauma or episode, see the next paragraph). If I finish both the phone-screen and interview phases with the Yes candidates and discover that none of those candidates is appropriate for the job, I continue the process with the Maybe candidates.

I send No candidates a written note or convey the information in a telephone call. However you choose to notify them, don't let too much time go by without letting them know your decision. Make the communication brief: "Thank you for submitting your application and résumé to our company. Unfortunately, your experience does not match our current needs, but we greatly appreciate your interest. We wish you success in your job search." This message can be delivered as a written note or verbally. I generally prefer to telephone the No candidates, speaking either directly to them or to the recruiter who sent me the candidate's résumé. If I am asked, I explain in general terms why the individual's experience is not a match for the particular job, but I am very careful to steer clear of statements that can be disputed or that could be interpreted as a subjective opinion rather than a fact.

Inform candidates of your decision as soon as you have made it.

Never leave a candidate wondering what you think about his or her chances. Even if you don't hire the candidate, he or she could be a source of candidate referrals for you. You can send e-mail, a letter, or a postcard, or you can telephone the candidate. If you received the résumé through an external recruiter, let the recruiter know the decision and why, and verify that the recruiter will contact the candidate.

When you first start reviewing résumés, you may need to take a couple of minutes per résumé to see whether the candidate has the expertise you require. With more practice in reading résumés, you probably can reduce your reading time to a minute or less per résumé.

Review résumés quickly. If you can't review a résumé and respond within 24 hours, make sure you or someone in your company acknowledges the résumé. Even if you only send an acknowledging e-mail or postcard, the candidate will know you've received the résumé. Don't sour a candidate on an opening just because you're in the midst of a crisis and can't take time to acknowledge the submission.

I try to make a decision within 24 hours of receiving a résumé about whether I want to proceed with the particular candidate. There are many reasons for responding to candidates quickly:

- If you're in a tight job market, you'll lose great candidates if you take too long reviewing their résumés.
- If you're repeatedly slow to respond to résumés, you'll develop a reputation as a dabbler, as someone who isn't ready to hire but just wants to see what's out there in the way of candidates. If you have this reputation, you will find it harder to attract great candidates when you do get serious about hiring.
- You'll establish a reputation as having a great company to work for, because you treat your candidates with respect.
- If you're working with recruiters, internal or external, they will be more likely to look for candidates for you, because you quickly let them know which candidate's résumés you liked, which ones you didn't like, and why.
- It's courteous to respond as quickly as you can to candidates. You'll feel better and so will they.

Look for patterns in your résumé-review process.

We all have prejudices and biases. When you review résumés, go back and look at the ones you rejected. Do you see any patterns in your rejections? Are you unknowingly biased for or against certain kinds of people?

Because of experiences you have had in the past, you may have developed a bias *for* or *against* people who are graduates of certain colleges, people from a specific region, or even people from another country. Maybe you're not comfortable hiring people who are older than yourself. Perhaps you've never worked with a man who you thought was sufficiently skilled as a manager, so you are wary of hiring a man for the open managerial position. Whatever the situation, be aware of your biases, and make sure you reject a résumé because the candidate's experience does not fit the open position, not because of gender, religion, nationality, race, and so on.

Make sure you don't screen out people who aren't "just like you." You probably don't need candidates who are clones of your current staff members, but a candidate who has had a different business experience from yours can bring added value to the job. For example, if you're looking for someone with real-time experience for a job in the computer hardware and circuit-board business, people who've worked on instrumentation projects in the automotive or aerospace industries may have both related experience and a valuable perspective to bring to your company, even though the industries they've worked in may be quite different from yours. Make your criteria as objective as possible, and actively seek out people who can bring different experiences, perspectives, and talents to your group.

You're human, so you, like me, probably will have subtle patterns linking which résumés you are more prone to accept or reject. That's okay. Learn what your patterns are, so you can choose whether to retain those choices, or make new ones.

Use résumés as feedback for evaluating your advertisements.

The résumés you receive give you information about the quality of your advertisements. Notice your reasons for putting a résumé in the Yes pile, the No pile, or the Maybe pile. How does your ad or job description encourage people with Yes qualities to apply? How could it better discourage people with No qualities?

Review résumés with a team to reach consensus.

One way to increase your résumé filter's accuracy is to review résumés as part of a three-person (or more) team. First, make multiple copies of the résumés so that each reviewer has a complete set. Then, have each person review all the résumés and sort them into Yes, Maybe, and No piles. Then re-group to discuss where people agreed and disagreed on candidates. As each person discusses how he or she made decisions, all interview-team members can learn how to review résumés faster and more effectively.[1]

---

[1] The specific point that people working in teams may be more insightful about evaluating potential employees than are individual reviewers is made well in Pierre Mornell, *45 Effective Ways for Hiring Smart: How to Predict Winners & Losers in the Incredibly Expensive People-Reading Game* (Berkeley, Calif.: Ten Speed Press, 1998), p. 37.

Case Study: Walker Software

As Dirk, the Development Manager at Walker Software, reviews résumés, he notes his reactions in the margin. His notes on Sally Developer's résumé follow:

617-555-ZZZZ   sallyd@home.com

Résumé

Sally Developer

17 Some Street

Cambridge, MA 02138

Okay, she's local.

Experience:

Jan. 1999 – May 2003: Machine Vision Company, Inc., Cambridge, MA. Software Engineer

Six years of experience. Appropriate.

She's worked on libraries, tools, and with QA. Hmm, also several apps.

- Designed and implemented machine-vision library with a set of common tools. Worked with two other engineers over an 8-month project.
- Led re-architecture effort for standard product line, based on new library. Re-architecture design effort took one full year, including building customer-usable upgrade modules.
- Worked with the QA technical lead on development of automated smoke tests. Part of a cross-functional team that developed and tested the smoke tests for use in daily builds.
- Instrumented library for performance testing. Took 8 months to instrument and improve performance.
- Developed several custom applications: box inspection, gauge inspection.
- Trained as a moderator for formal inspections (Mar. 2002).

Led architecture effort? With only six years of experience? I wonder what that was like.

Hmm, smoke tests and performance experience. Real-time, too. That's good.

May 1997 – Jan. 1999: Small Software Company, Inc., Cambridge, MA. Software Engineer

I don't care about hobbies.

- Developed server applications for Small Software's flagship client/server product. Included Oracle 8.0, Sybase 10.0.2 stored procedure development.
- Ported server applications from UNIX to NT.

Received degrees while working. Shows drive, initiative, and focus.

Good academic background. I know other people from BU. They were pretty good.

Hobbies: Rowing, game development, bicycling.

Education:

M.S. Software Engineering, Boston University, Boston, MA, January 2001.

B.S. Computer Science, Boston University, Boston, MA, May 1997.

References furnished upon request.

Dirk's notes tell us what he reads between the lines when reviewing Sally's résumé. First, he verifies that the candidate is local—an important detail because Walker is not paying for candidate relocation. Then, he calculates years of experience (six) by subtracting the date of the candidate's undergraduate degree (1997) from the current year (2003). The candidate hasn't included much about each of the projects, so Dirk notes that he can't tell what positions she's had and the duration of each project. He is interested in the candidate's work with other groups, her real-time experience, and her performance experience. Dirk suspects that if she helped develop smoke tests, maybe she knows enough to read other people's code. Sally's hobbies indicate that she is not a big risk-taker or thrill-seeker and that she likes intellectual challenge as well as physical activity, but as they are not relevant to the job, Dirk doesn't care what they are. Dirk decides to place this candidate's résumé in the Yes pile, primarily because she appears to have enough appropriate experience to be worth phone-screening.

POINTS TO REMEMBER

- Review résumés using a loose filter for maximum hiring flexibility. Especially if you're hiring under a deadline, it's worth your time to phone-screen more candidates rather than fewer. You'll spend more time phone-screening, but less time interviewing.
- Look for the context in which the candidate performed prior work. How similar is that context to your context? Does similarity of context matter?
- Review résumés quickly, and respond to candidates as quickly as possible with your decision, even if your decision is Maybe or No.

Part 3

Preparing to Interview Candidates

If you're lucky, you have a fabulous organization behind you, which helps you to analyze the job, write the job description, recruit candidates, and review résumés. All you have to do is interview. If you're not so lucky, you've performed all the preparation work yourself.

Interviewing can be fun for everyone concerned, as long as you know how to ask questions, and how to plan both phone and in-person interviews. I use a table technique to measure interviewing success, which you may want to adapt if you'd like to track your success in this area:

Week of	# of candidates phone-screened	# of candidates interviewed in the first round of phone-screens	# of candidates interviewed in the second round of phone-screens	# of candidates deemed unqualified and eliminated during the two rounds of phone-screens	% of candidates deemed unqualified and eliminated after the first round of in-person interviews	# of offers made to first-round candidates	% of offers made to second-round candidates

During a hiring search, you will phone-screen more candidates than you will interview in person. First, I track the date (Week of). In the next column, I keep track of the total number of phone-screened candidates. Some candidates will be eliminated after the first phone interview, but others may be scheduled for a second phone interview. I like to keep track of these numbers, so I can detect if my phone-screen questions are adequate. If I need more than one phone-screen, perhaps my filtering and, especially, my elimination questions are not adequate. And, if you have multiple rounds of in-person interviews, some candidates will not move to the second round.

After your first interview with a candidate, decide whether the candidate fits the job or not. It's better to find out when you phone-screen people that they don't fit the job than to find out during an in-person interview because less valuable time is wasted. Track your percentage of poor-fit in-person interviews, in order to learn how tight your résumé-review filters are and how good your phone-screen questions are. The tighter your résumé-screening filters or the better your phone-screen questions, the fewer poor-fit or unqualified candidates you'll interview. To reduce the number of such time-wasting interviews, try to improve your phone-screen questions. Talking to candidates during the phone-screen will give you a better indication of their abilities than just reading their résumés.

You won't offer jobs to all the candidates, so track the number of offers you make. If you're phone-screening more than 25 candidates to every interview you set up, maybe your résumé filters are too loose. If you find that you have 20 percent or greater poor-fit in-person interviews, check your phone-screen questions to detect what's not working well.

Review the number of offers you make. If you're satisfied with the ratio of offers to candidates, great. If you think you're interviewing too many people in person, then review your phone-screen questions. I aim for an offer ratio of at least 20 percent; that is, my goal is to offer a job to at least one out of every five second-round candidates. Your ratio will vary based on your criteria for the first-round and second-round interviews.

If you aren't interviewing many candidates, look back at your recruiting activities and phone-screens. Then, ask yourself whether you receive enough résumés. If you receive enough of what seem to be good résumés, look again at your phone-screen questions. If you're phone-screening too tightly, you won't interview many people. If you don't receive what you consider to be a sufficient number of résumés, it may mean that you're not advertising sufficiently so that people know about your job. If the wrong people are answering the ad, rewrite the ad to discourage candidates who are not qualified, while encouraging the suitable candidates.

The four chapters in Part 3 deal with interviewing skills, how to use those interviewing skills in phone-screens and in the in-person interview, and how to follow up after the interview. Chapter 7, "Developing Interview Questions and Techniques," will help you prior to the interview. Chapter 8, "Creating and Using Phone-Screens," will help you apply interviewing techniques to phone-screens. Chapter 9, "Planning and Conducting the In-Person Interview," will help you prepare for and carry out first- and second-round interviews. And, Chapter 10, "Following Up After the Interview," will help you work with your interview team to assess candidates after the interview.

7

Developing Interview Questions and Techniques

Stanley: "Hey, Joe—you interviewing the performance tools candidate today?"

Joe: "Yeah."

Stanley: "What are you going to talk to this guy about?"

Joe: "Oh, I don't know. Whatever I feel like, I guess. How the heck should I know what to ask a performance tools guy?"

If you were ever concerned that you didn't know what to ask in an interview, or didn't feel as if you were successful in selecting questions to ask in an interview, this chapter will help you determine what techniques to use for interviewing and what questions to ask, and how.

Whether you're phone-screening a candidate or conducting an in-person interview, choose your interview questions carefully to make the most of your time with your candidate. The interview can help you learn specifics about the candidate if you ask questions that answer the following:

- What functional skills, product-domain expertise, tool and technology skills, and industry expertise does this candidate have?
- What personal qualities, preferences, and technical and non-technical skills does the candidate have?
- Do the candidate's personal qualities, preferences, and skills match the job's requirements for technical work and the organization's culture?

As you listen to the candidate's responses, you will gather a general impression that can help answer one additional question: Can you trust the candidate's answers?

Design your interview questions and activities to learn the answers to the preceding questions. Whether your interview is a phone-screen or an in-person interview, choose how you'll ask which kinds of questions. You can ask many of the same questions in a phone-screen as you would in an in-person interview, but if you plan to both phone-screen and interview candidates, change your in-person interview questions so that you broaden the discovery process.

Choose which kinds of questions to ask.

By asking great interview questions, you not only learn what you need to know about the candidate, but you also help sell the candidate on your company, on your group, and on yourself, either as a member of the organization or as the group's manager, or both. Interviews are too short to waste time asking questions that don't elicit information about how the candidate will work in your organization.

You have multiples types of interview questions from which to choose: close-ended, open-ended, hypothetical, audition, and meta-questions. With a little planning and thought, you'll be able to choose your interview questions to assess the candidate's overall suitability for your job and your organization.

In my opinion, the most effective interview consists of a series of close-ended questions to establish the basic facts, used in combination with open-ended, behavior-description questions, followed by auditions. However, there are additional question types you may find useful, such as hypothetical questions or meta-questions—questions about the question.

Close-ended questions

A close-ended question helps you verify facts. The candidate answers a close-ended question with a brief, factual response. Use close-ended questions such as the following to establish the facts: Where are you working now? What is your salary? How many years of C experience do you have?

Close-ended questions create boundaries around the open-ended questions you'll be asking. I use close-ended questions to establish how long a person has worked with a specific technology or in a particular industry. You may think the candidate's résumé establishes the answer to these questions, but candidates sometimes stretch the truth or use phrasing on a résumé that may be misleading. I once interviewed a candidate who had included the following phrase on his résumé: "Eight years experience in C++ and UNIX." During the interview, I asked, "How many years of C++ experience do you have?" His response, "Three." I then asked, "And how many years of UNIX experience do you have?" "Five," came the response. "Ah! So, on your résumé where you listed eight years, that was because you added the years together?" "Yes."

Candidates generally will answer close-ended questions accurately during a phone-screen or in an in-person interview.

Establish the facts early in the interview, whether the interview is a phone-screen or an in-person interview. That way, you can move quickly to open-ended questions.

Open-ended questions

Open-ended questions help you probe for more information than just the facts. They require more than a yes or no answer. One generic, open-ended question I like to ask is, "Tell me about your current job."

Open-ended questions help you learn more about the candidate, but they don't focus on specifics. Use behavior-description questions to focus on a particular area or experience.

Behavior-description questions

Behavior-description questions, a form of open-ended question, help you discover how a candidate has worked in the past. If you know about behavior-description interviewing, you have probably read that "the best predictor of future behavior/performance is past behavior/performance *in similar circumstances.*"[1] Two corollaries are helpful in evaluating behavior consistency:

1. The more recent the past behavior, the greater its predictive power.
2. The more long-standing the behavior, the greater its predictive power.[2]

Although people do learn new technical skills as well as how to apply those skills, they rarely change behavior. When managers give me reasons for why they have fired someone, they typically say things such as the following:

---

[1] Tom Janz, Lowell Hellervik, and David C. Gilmore, *Behavior Description Interviewing* (Englewood Cliffs, N.J.: Prentice-Hall, 1986), p. 32.

[2] Ibid. op. cit., p. 33.

- Sam didn't have enough drive.
- Sally wasn't adaptable enough.
- Steve didn't come to work on time.

Each person's behavior was a key factor in why he or she was fired. When a manager says, "Sam didn't have enough drive," the manager is speaking about initiative. A candidate will indirectly tell you about his or her initiative in answering an open-ended, behavior-description question. You can ask, "Did you ever notice problems at your company that you would have liked to see fixed?" If the candidate answers "Not really," then probe further to see whether things really did run smoothly, or whether the candidate is too polite to answer the question with a yes. If the candidate answers "Yes," then continue the dialogue: "Tell me about what you would like to have seen changed." Follow up that discussion with, "Were you able to effect any changes, and if so, what were they?"

You can detect initiative, adaptability, punctuality, or any of the candidate's personal preferences, qualities, and skills if you ask behavior-description questions and compare the circumstances of the behavior to your environment. By weighing the candidate's behavior along with his or her technical expertise, you can fairly confidently predict his or her probability of success in your organization.

I generally start an interview by asking the candidate to tell me what he or she is currently working on, or, if the candidate is unemployed, I ask for a description of the most recent project. I tailor my question to elicit information relevant to the functional area for which I'm interviewing. For example, for a developer, I'll say, "Tell me what you're developing now." For a tester, I'll say, "Tell me about your test choices and activities now."

If that's too open-ended an approach for you or the candidate, other questions, which can be tailored to the type of candidate you're interviewing, are given below, in Table 7-1. When you ask a series of questions or ask questions that have multiple parts, pause and wait for the answer between each question or part. Whether you're asking questions on the phone or in person, you probably will want to take notes, and your silence as you take notes may prompt some people to add detail to their answers to you—a very positive benefit.

If you find it difficult to draw out a candidate enough to learn anything of substance, you may be more successful in getting the candidate to open up if you use hypothetical questions.

Open-Job Title	Information Sought	Questions to Ask or Tailor
Developer	To learn how the developer approaches design or implementation	Tell me about the piece of the project you're working on now. Tell me how you came up with that design. What issues did you run into during implementation? Can you tell me about a time when you had trouble designing a product? Can you tell me about a time when you had fun in the design/implementation/debug phase? What design techniques do you use, and what makes you choose those techniques?
Tester	To learn about the tester's functional test skills, and possibly about how the tester plans	How did you decide what to test in your project? What test techniques are you using for which pieces of the project? If you hear an answer like, "My boss told me what to do," then follow up with any of these questions: • How are you testing the product? • How do you track which tests pass and fail? • How do you know when you've found a problem? • Tell me more about how you report problems. • Tell me about a great defect you found.
Support Representative	To learn how the rep manages daily work	Tell me how you record an incident. What are the first, second, and third things you do after receiving an incident report? Tell me how you've handled a difficult customer. Tell me about a time when you were particularly proud of something you did for a customer.
Writer	To learn how the writer determines writing requirements and how he or she elicits information to know what to write about	Tell me about your current writing assignment. How did you determine the requirements for the writing part of the project? How did you know what to write?
Project Manager	To learn how the project manager approaches projects, and about previous project management experiences	Tell me about how you're managing your current project. What's similar about this project? What's different about this project from others you've managed?
Technical Lead	To learn how the candidate provides technical leadership	When things are proceeding well on the project, what do you do? How do you recognize when the work is proceeding well? How do you know when things aren't proceeding smoothly, and what do you do?
Manager	To learn about how the manager performs daily management assignments	Give me an example of how you manage difficult people. Give me an example of how you manage peer relationships. Give me an example of how you give feedback and evaluate employee performance.
Technical Staff Member	To learn about how the person interacts with other people and how he or she perceives the jobs	How well do you receive feedback about your own work? What's fun about your job now? What's challenging for you in or outside of your job? What's difficult for you and why?

Table 7-1: Job-Title-Related Questions for Interviewers (continued on next page).

All candidates	To learn about the person's initiative	Tell me about a time you didn't have enough information to complete an assignment, and explain what you did.
All candidates	To learn about the person's sense of responsibility	Have you come across any problems waiting to happen, and if so, what did you do?
		How do you track your progress on your current work project?
		How is that tracking method similar to or different from what you used on your previous project?
		How do you estimate your work on your current project, and how is that estimation method similar to or different from that used on your previous project?
All candidates	To learn about the person's drive or ambition	When was the last time you wanted a promotion or a particular assignment?
		What did you do to obtain what you wanted?

Table 7-1: Job-Title-Related Questions for Interviewers (concluded).

Hypothetical questions

A hypothetical question can draw out information because it asks how someone would behave in a specific type of situation. Hypothetical questions also can be useful when used in addition to open-ended, behavior-description questions. For example, when you're trying to determine what a person takes into consideration when solving problems, you could ask, "What would you do in *x* situation?" Listen more for clues about the candidate's thought process than to the answer. What people answer to hypothetical questions is something they would like to do, not something they've ever done. They can't know how they'll react unless they've been in the situation.

If you have hypothetical questions that have served you well and address people's work behavior, keep those questions in your interview repertoire. However, don't bother asking hypothetical questions that aren't specifically relevant to the work to be done. Replace a question such as, "If you had two wishes, what would they be?" with a question more likely to elicit information about the candidate's behavior on the job. For example, if you were interviewing a technical support engineer, you might ask, "Have you ever had to choose between supporting a customer and reporting status to your management?" If the candidate says "No," ask, "What would you do if you had to choose, and why?"

Unless you link the hypothetical question to work behavior, the answer won't tell you anything relevant about the candidate. Too often, candidates tell you what they think the right answer is, rather than what they really believe. One development manager I consulted with told me he had asked a candidate, "How would you react in a release meeting where the developers want to release and the testers don't want to release?" and the candidate, suspecting that the development manager would be biased toward developers, answered, "I'd agree with the developers." As a result, the development manager didn't discover that the candidate had been in just that type of situation the previous month and had facilitated a reasonable approach to the release. If the manager had asked, "Have you ever been in a position where the developers wanted to release and the testers didn't?" the candidate honestly could have answered "Yes," to which the follow-up question "What happened?" or "What did you do?" would have encouraged a healthy dialogue.

Answers to hypothetical questions can help you evaluate a candidate's probable reactions to events and can give you insight into the way the candidate works. I once interviewed a tester after she'd completed a project she called "dissatisfying." I asked, "What made the project dissatisfying?" She replied, "I never did get a handle on the testing. I was always reacting, not planning, not looking into new areas." "Did that ever happen to you before?" I asked. "Well, not like this," she said. I then asked, "Do you know what you'll do differently next time?" In response, she described a series of actions she would take the next time that would help her plan and assess where she was at each stage of the project. Because I believed her answer indicated that she was someone who learned from past mistakes, I hired her and coached her through behavior changes for the next couple of projects. Had I not asked what she'd do differently next time—making her imagine a hypothetical situation—I would not have known her potential. She proved to be well worth the risk.

Schedule auditions to allow candidates time to demonstrate their abilities.

Auditions provide the hiring manager and screening team with a way to watch a candidate in action—and they work well for hiring both managers and technical staff.[3] To get the most out of the audition, I ask candidates to envision themselves in a scenario or problem situation I describe. Specifically, I ask the candidate to take some time to solve the problem, and then ask him or her to walk through the solution with the hiring team and me. (You can conduct auditions with the entire hiring team or one-on-one with

---

[3] For details on how auditions can be structured, see Tom DeMarco and Timothy Lister, *Peopleware: Productive Projects and Teams,* 2nd ed. (New York: Dorset House Publishing, 1999), pp. 103ff., and Gerald M. Weinberg, James Bach, and Naomi Karten, eds., *Amplifying Your Effectiveness: Collected Essays* (New York: Dorset House Publishing, 2000), pp. 59ff.

one interviewer. The more people who see the audition, the better your evaluation of the candidate's audition will be.)

The key to a successful audition is to give the candidate enough time to prepare—to explore the product or prepare the presentation or investigate a problem—so that he or she can talk intelligently about the problem. Keep in mind that a scenario posed in the form of a game, puzzle, or non-work-related problem-solving task does not constitute a credible audition; such activities are not worthwhile primarily because they are non-work-related assignments.

Developer auditions

When auditioning developers, I use several audition approaches:

- **Debugging:** I show the candidate a software product in development, including a piece of the code, and ask him or her to solve a problem that I tell the candidate already exists in the code. (In this audition, I intend to see how the person looks for problems, isolates problems, considers solutions, looks for alternative solutions, chooses a solution, and implements a solution. If you will have candidates look at production code, be sure to carefully select the code sample in advance of the interview. Rather than having the candidate work on a computer or at a workstation, print out the selected segment of code so the candidate doesn't possibly see something proprietary that you don't want seen.)
- **Product design:** I explain a desired addition or modification to one area in a particular product's design. I ask the candidate first to re-design to satisfy my request, and then to explain why he or she chose the particular approach. (In this audition, I'm looking at how candidates examine designs, what issues they are sensitive to, and how they know either that the design is complete or what they would have to know in order to determine that the design is done. Be sure to select the product area in advance, so you know what to discuss with the candidate.)
- **Product review:** I give candidates the work product (either design specifications or actual code), and ask them to read the material and mark any areas about which they have questions or issues. Then, to determine their proficiency and analytical skills, I ask them to explain their questions and issues.

- **Programming and development:** I define the requirements for a small program (random number generators or bubble sorts are good examples), and then ask the candidate to write the code. (It is especially critical in this situation that you give the candidate enough time to complete the audition.) When I turn the code produced in this audition over for review by a programmer on the hiring company's staff, I ask him or her to look at whether the candidate has properly declared functions and objects, initialized variables, managed pointers or memory (such as allocating and freeing memory), or handled whatever points are important to the job.[4]

Another form of audition is to request that the candidate bring in samples of his or her work, especially if that work is open source. Although not everyone can bring in work samples, the more evidence of how a person works, the better your evaluation of the candidate.

Use your own company's products for the problem-solving auditions to avoid violating any candidate's nondisclosure agreement with his or her current or previous employers. Try always to use pieces of your company's design or code that could not in any way be construed to be a closely held secret, but if you must disclose proprietary matter, remember to have the candidate sign a nondisclosure agreement.

Tester auditions

One good technique for auditioning testers requires a product demonstration. I provide a brief written description of a part of the product and then give the candidate tester some time with the product to explore it. I then have one of two choices: Either I ask the candidate tester to walk through his or her thought process on how to test the product or I ask the candidate to perform actual testing and problem reporting. In cases in which it is possible to show the candidate the running product rather than just a description, I sit the person down in front of the computer, and say, "Go test. Here's some paper so you can take notes. I want you to walk me through your testing in twenty minutes." You can look for not just the person's specific approach, but also how he or she went about exploring the product.

If I have enough time in the audition time slot, I ask the tester to perform tests and write problem reports. In this case, I don't intentionally

---

[4] A classic discussion regarding testing the skills of programmers is Gerald M. Weinberg, *The Psychology of Computer Programming: Silver Anniversary Edition* (New York: Dorset House Publishing, 1998), pp. 175ff.

bebug (insert known problems into) the software—unless the product is mature and stable. If the product is still in development, I tell the candidate that he or she should expect to find problems.

If I am looking for both exploratory testers and test planners, I may decide to use both techniques (walking through the product, and actual testing) but in two separate auditions. When looking for graphic user interface testers, I recreate part of the product on paper and have tester candidates walk me through how they would test from the GUI.

Senior staff auditions

Asking candidates who apply for senior-level staff positions to give a presentation about their current project or research seems worthwhile only if they'll be called on to make presentations about their work in your organization. Otherwise, you're asking them to perform the work of creating and presenting technical information that isn't relevant to the position for which they are interviewing.

If you're going to ask candidates to make a presentation, make sure they know in advance that an audition of this type is part of the interview process, so they can prepare. And, if you do ask a candidate to prepare a presentation, make sure you have access to whatever equipment he or she may need, such as a laptop, a projector and screen, a blackboard or white board, flip charts and pointers, and chalk and markers. In addition, let the candidate know how much time you will allocate for the presentation.

Project manager auditions

For project management candidates, you can show sample project documents and ask them to point out areas they think might entail risk. Or, show candidates a sample project schedule and ask them to describe what they would consider to be the critical path; ask them to raise any questions they have about the critical path. Ask them whether they have worked on projects similar to the sample; if they have had similar experience, ask for details about what they did and when. If you expect your project managers to run project meetings, have candidates develop an agenda for a project team meeting. If you're looking for a project manager with excellent facilitation skills, have candidates facilitate a real project team meeting.

A type of project management audition I like to use when I'm looking for a manager for a high-visibility or large project requires candidates to thoroughly describe a project they have managed. I ask the candidate to describe how he or she organized the particular project and what the deliverables were, and then I ask the candidate to describe what variation on that organization he or she would use according to several lifecycles, such as staged delivery, SCRUM, spiral, or waterfall lifecycles. Depending on how familiar the candidate is with various lifecycles, I may then ask which lifecycle he or she would choose if doing the project again—and why—and what impact such a lifecycle might have on the interim and final project deliverables.

Manager auditions

For managers, I sometimes set up an audition to take place during a lunch-time meeting, and I invite attendees from the group with which the candidate would be working if hired. The audition segment could consist of a brief presentation by the candidate, to be followed by a Q & A session. I note to whom the candidate talks most frequently, and whom the candidate avoids.

If you're concerned whether a candidate will be able to manage the project portfolio, your audition might consist of asking him or her to organize the portfolio and then present the selected organization to your group.

Two management skills that many managers find troublesome are providing effective feedback and assigning work. To test the candidate's ability to actually perform these management tasks, first ask behavior-description questions to establish that the candidate has in fact attempted to master these skills. Then, provide the candidate with a hypothetical scenario where he or she needs to assign work to group members. In addition, provide the candidate with a fictional group member's work and ask the candidate to provide feedback in a role-play.

Tailored auditions

As you develop auditions, consider your product context and derive the audition from the product. For example, if you develop a work-flow application, you could request a candidate developer to speed up the application, a candidate tester to test for improved speed, a candidate writer to document how to use the improved speed, or a candidate technical support rep to determine where the application is still too slow.

Formulate a set of meta-questions.

Asking questions about the answers you've already received can help you narrow the candidate pool. I call these questions meta-questions because I use them to draw out details that add a layer above the information I've already gathered. Some possible meta-questions follow:

- "What haven't I asked you to talk about that I should have?"
- "If I were to ask you a question for which an honest answer would make you look less qualified for this job, how would you answer?" (This question is a variation on the question, "What are your strengths and weaknesses?")

Meta-questions can help you regain focus when an interview isn't developing as you had expected. I used a meta-question not long ago to find out why a candidate who had seemed outstanding during the phone-screen appeared to flounder when I brought her in to meet with me in person. After talking with her for fifteen minutes and detecting some confusion in details about her prior managerial experience, I asked, "Is there something I should have asked you during the phone-screen?" She answered yes, and explained that because I hadn't asked about her *direct responsibility* for the work we had previously discussed, she may have misled me. She went on to explain that although she had general management experience, she felt she didn't know enough about the functional area for which we were interviewing her. My meta-question brought out her admission, and helped me determine that she was not qualified for the particular job. For subsequent interviews for the position, I modified my phone-screen questions either to ask about specific experience in the functional area or to ask the meta-question itself.

Although I didn't discover that disqualifier during the phone-screen, I was able to see a difference between the way the person had responded on the phone and the way she responded when she sat across the table from me. The meta-question provided the reason why.

Meta-questions can also be helpful at the end of the interview, once rapport has been established with the candidate. I frequently ask, "What haven't I asked you yet?" toward the end of the interview—and uncover information that helps me with my hiring decision.

Learn to avoid asking irrelevant questions.

Irrelevant questions focus on the *person* rather than on the person's *qualifications* for the open job. Interviewers sometimes use irrelevant questions to draw out the person behind the bare facts listed on the résumé. Questions such as "Where do you want to be in five years?" or "If you could change just one thing, what would it be?" are irrelevant because they do not directly relate to the qualifications required for the job. Answers to such questions may satisfy your curiosity as an interviewer, but the answer is irrelevant to the job you need filled. Ask yourself whether you really care where the candidate wants to be in five years. If the candidate answers truthfully, "I'd like to be backpacking in Nepal," or "I want your job, you idiot," what have you learned about the candidate that applies to the job you need filled? Nothing. And for those of you who think that "I want your job" is an answer, think again. Candidates may think this is a leading question and that you want someone ambitious. Candidates may think you will value them more if they appear ambitious, whether they want your job or not.

In addition, if you use leading irrelevant questions such as "You don't like projects that fail, do you?" you're answering the question yourself, instead of asking the candidate what he or she has done when a project looks as if it's in danger of failing.

Irrelevant questions don't help you understand how a candidate works, so why would you even bother asking? Asking candidates where they want to be in five years when the project is scheduled for six months and the company is in danger of losing funding in two years, seems silly to me.

I avoid asking irrelevant questions in interviews primarily because I've found that many candidates say what they think sounds good to the interviewer, regardless of what they really believe. If I want to know how ambitious a candidate is, I'll ask a behavioral question such as, "When was the last time you decided you wanted a promotion, and what did you do to obtain it?" If you use irrelevant questions in interviews, reconsider what you're asking and why. As an alternative, use behavioral, open-ended questions during at least part of your interview time, and auditions for another part. You'll come away with a more complete and revealing picture of the candidate.

Candidates react in various ways to irrelevant questions. In one interview, as soon as I had introduced myself, the candidate took charge of the interview, saying, "Look, I don't want to solve world peace. I don't know where I want to be in five years. I couldn't care less about solving some stupid game while you time me. If you don't want to talk about developing software, I'm done with this interview." I told him we would indeed be talking about developing software, and he appeared relieved and possibly even thrilled at the end of our behavior-description-oriented interview. But, did I hire him? Well, no—because while I appreciated his honesty and his straightforward approach, I determined he lacked the patience and social grace required for that particular open position. I would have hired him had a large degree of patience and a modicum of people skills not been essential.

Combine question types to make the best use of available time.

Review your job analysis worksheet to decide which areas you'll ask which kinds of questions about, combining open-ended, close-ended, behavior-description, hypothetical, meta, and so on, as needed. For example, following an audition, you can use hypothetical questions to draw out more about what a candidate might not have had time to perform in the audition. Ask the candidate questions such as, "Where else would you look to simplify the design?" or "Where else would you look for problems in the software?" Meta-questions can also be useful as the audition nears an end: "What other tasks should I ask you to perform?"

When you're interviewing a candidate, it's natural to start off with close-ended questions to establish the facts and follow up with behavior-description questions to understand how the candidate has worked. In addition, you can follow the behavior-description questions with meta-questions or hypothetical questions to see what candidates have learned and what they plan to do the same or differently the next time. Once you've learned about the work a candidate performed, consider some meta-questions or hypothetical questions such as, "How was that work for you? If you had a similar project, what would you do differently? Similarly? What lessons did you learn?"

In an interview with a candidate who had managed a particularly stressful project a few years earlier, I asked, "How was that project for you?" The candidate paused, collecting his thoughts, and then said, "You know, I'm not sure I ever want to work on a project like that again. I'm not going to take on impossible projects. But talking about it with you just now makes me think of some questions I need to ask you: Do *you* ask your project managers to take on impossible projects? And, if you say you don't, how do I know that you really don't?"

With this concern aired, we were able to discuss what made a project impossible for that project manager and what made a project impossible for me. The frank discussion that ensued enabled us to better understand each other's values and professional requirements and convinced me to hire him. His was—and still is—one of my most successful placements.

I've found that I use my interview time best when I ask close-ended questions to establish the facts; follow up with behavior-oriented, open-ended questions; and then end with an audition. However, because many candidates need time to think through their answers, I must use my limited interview time wisely, taking care not to allow the interview to degenerate into a rapid-fire interrogation. I try to make the interview a congenial conversation, while still sticking to questions I have determined in advance. I limit the number of areas I ask about in a phone-screen, and follow the same approach for an in-person interview. (For more information on how to choose questions, see the next two chapters.)

Following are the kind of open-ended, behavior-oriented questions—and my thought process as I ask them—I could use when hiring a test manager to work in a start-up environment:

> "Do you ever have to release products before you've been able to perform enough testing?" If the candidate says no, I ask, "How would you go about determining when enough testing has been done?" If the candidate says yes, I ask, "How do you decide what to do?" With these questions, I'm hoping to draw out an answer that reveals how the candidate makes tradeoffs and how he or she works with other product-development managers. If the candidate hasn't discussed tradeoffs with peer product-development managers, I need to find out whether he or she has the experience I believe will be required to manage the testing function in the hiring organization.

When I interview technical-support people, I want to learn how they have dealt with difficult people. One line of questioning I take to get the information combines a bit of humor with irony:

> "We sometimes deal with difficult customers," I say, usually with a wry grin. "Tell me about a customer you've dealt with that was difficult, and then tell me what you did to manage the situation." Every technical support representative in the universe has dealt with at least one such frustrating situation. When I ask this question, I gain insight into the kinds of people or situations the candidates thinks are difficult, and how the candidate handles such encounters. I learn how the candidate manages his or her own emotional reactions, how he or she negotiates with customers, and possibly even how he or she will vent to peers after an encounter.

When I interview senior-level technical managers for a position, I try to learn how they approach the problems of staffing, including personality clashes, funding, and overtime. The candidate's answer is especially important for projects on which salaries constitute the largest part of the budget. I typically start my questioning as follows:

"How do you know when you have the right staffing level?" I listen for answers that analyze the needs of the business and indicate that the person understands that when the business environment changes, staffing needs change. Sometimes candidates answer this inquiry in a close-ended way with a brief, uninformative answer: "When the projects are staffed." Although this seems somewhat like a wise-guy's answer, it does reflect what has become a predominant way of thinking among senior-level managers—that the correct number of people somehow correlates with the needs of a specific project. When I hear such a comment, I ask how the candidate knows he or she has the right people on board. As in a conversation with a toddler who asks questions for the sake of asking more questions, this line of questioning can become counterproductive, but it also can be surprisingly informative. It's up to you to steer its direction toward information and away from silliness.

Part of what I look for in the answers is information that will tell me how the candidate's current job environment is similar to the one I am trying to fill and how it's different. If the candidate currently manages three people and I'm looking for a manager for a twenty-person department, I listen for indications of how the candidate takes the environment into account—if the candidate typically describes only one possible solution for a given problem, he or she may not be able to manage a larger department effectively. When I am evaluating candidates who want to increase their responsibilities, I look for adaptability and the candidate's ideas about which solutions are appropriate for which context.

Especially in limited-time interviews, whether they are phone-screens or in-person interviews, behavior-description questions help you extract useful information about a candidate. A candidate's answers to these questions will give you a better idea of whether or not the individual will fit into your organization and be successful there.

Ask all candidates applying for one position the same set of questions.

For an open position in a large company, I recommend that candidates be interviewed by multiple interviewers, all of whom have been trained to function as parts of a single interview team, in a series of one-on-one interview sessions. For small companies, I recommend that solo interviews be combined with a group event, such as a lunch, because it's important for candidates for jobs in small groups to be able to develop rapport with the rest of the team. Regardless of the company size, however, the interviewers should coordinate their questions with other interviewers so that each applicant gets asked the same set of questions as all other applicants for the position. For more detail on how to organize interview-team members and their questions, see Chapter 9, "Planning and Conducting the In-Person Interview."

Since not everyone should ask about all of the qualities, preferences, or skills in an interview, each member of the interview team should take on a couple of areas to ask about in more detail. Each interviewer needs to tailor his or her set of questions to draw out information to reveal a particular work ethic or a specific talent. Interviewers must coordinate with other interviewers so that all aspects are covered without redundancy, or they should be assigned questions to ask. I develop at least two open-ended, behavioral questions for the specific areas I'll ask about in the interview. For example, if I am looking for someone who exhibits good problem-solving skills, I ask for detail, saying, "Give me an example of a time when you ran into a problem. Tell me how you resolved it." Keyed to specific job types, the questions given in Table 7-2 can help you draw out information about a candidate's problem-solving skills.

Job Title	Questions to Reveal Problem-Solving Skills	Related Questions
Developer	Tell me about a design problem you had.	What was the nature of the design problem? How did you fix it? Did you gather data, get help from other people, or think it through? How long did it take you to solve the problem? How did you know the problem was fixed?
Performance Developer	Tell me about a time when you were concerned about performance.	Did you gather or evaluate data for measuring performance? Did you conduct other forms of testing or peer reviews? Did you develop a process for changing the product?
Support Staff Member	Tell me what happens when you get a phone call from someone who's stumped.	How did you go about obtaining information about the product? How did you go about helping the user to describe the problem? Did you develop a systematic approach to problem definition?
Test Staff Member	Tell me what you would do to test the product without knowing what's in the product.	What do you do to learn about the product? How do you go about getting the product definition? How do you extract information from whom/what/when? When you test the product with knowledge, how do you know the definition is complete?
Technical Writer	Tell me about a time when you had a hard-and-fast deadline but got stuck writing.	What did you do about the deadline? What do you do to get past writer's block? What have you written about a specific release or product you've used?

Table 7-2: Assessing Problem-Solving Skills.

Ask questions to reveal cultural fit.

Not only should you ask questions about the job, you also should ask questions that will enable you to get an understanding of how well the candidate will fit into your company. Look at your cultural-fit and elimination factors, and consider questions such as the ones in Table 7-3:

Category	Questions to Reveal Cultural Fit	Related Questions
Overtime	Tell me about any experience you have had with working overtime at the end of a release.	Have you juggled overtime with other commitments? Would you be available to put in overtime on an infrequent basis? On a long-term basis? Never?
Travel	Tell me about your experience with assignments requiring travel.	How have you handled travel in the past? Could you travel half of the time during April through June and in October and November?
Extracurricular and Volunteer Activities	Tell me about any work you've done on a volunteer basis and whether you'd be amenable to volunteer work.	Have you participated as a volunteer on a one-time basis or done extensive service? Would you be amenable to volunteering regularly to the United Way or a like charitable, fund-raising campaign?
Training	Tell me about your previous experience with company training policies.	Have you ever determined your own training policy? If so, what was the curriculum? If not, what kinds of training are you looking for? What training have you had that addresses the functional skills, tools, or industry for the products you have worked on previously? What's the value of that training to you?

Table 7-3: Assessing Cultural Fit.

Ask contractors the same questions you ask prospective staff hires.

I ask many of the same questions when interviewing a contractor as I use to interview a candidate for a permanent staff position. In both cases, I need to understand the candidate's abilities and skills in enough detail to know whether he or she will be successful in the position. One question I make sure I ask contractors is, "How do you handle completing a contract and what actions do you take for the hand-off?" I want to know what preparation a contractor does prior to hand-off, and what deliverables were required. Specifically, I want to avoid contractors for whom the hand-off becomes a last-day affair.

If you work with a contractor whom you want to convert to employee status, take the time to develop and ask questions specifically about cultural fit and any requirements the contractor has for time off.

Help non-technical interview-team members develop questions in their own area of expertise.

People on the interview team who are not members of the technical staff can ask questions pertaining to their own area of expertise. So long as someone technical is running the audition and asking questions about the technical skills, non-technical members of the interview team can provide a valuable contribution to the screening process. Chapter 9, "Planning and Conducting the In-Person Interview," provides more information on this topic, but the following instructions can be useful for the non-technical member of the interview team.

- Ask close-ended questions to establish facts about previous and current employment history, education, experience, and so on—information you will verify after the interview.
- Ask previously selected behavior-description questions to draw out details about the candidate's experiences and achievements.
- Ask whether the candidate has questions, and provide whatever answers you can, along with a time frame within which additional answers will be provided.
- Ask, "Is there anything else I should ask you?"
- Close the interview.
- Give the candidate breaks between interviewers to make a phone call, use the restroom, or get water, coffee, or the like.

POINTS TO REMEMBER

- Ask questions specifically worded to help you determine whether the candidate will fit your open position and culture.
- Use a variety of questions to elicit the most information in a short time.
- Spend your interview time primarily using behavior-description and audition questions.
- Develop multiple ways to ask about specific skills, preferences, or capabilities, framed within the context of the candidate's work.

8

Creating and Using Phone-Screens

Interviewer: "Oh, yes, let's see. How many years of Smalltalk experience do you have?"

Candidate: "I don't have any Smalltalk experience. Why?"

Interviewer: "Oh, too bad. We only need people who know Smalltalk. Hmm, maybe the job posting didn't make that clear, but it's the case. So, I guess there's not much point in continuing this interview. Thanks for coming in, but we really can't use you."

From the preceding, we can imagine a scene in which an interviewer sitting at a desk is reading a candidate's résumé for the first time. The candidate, seated opposite him, fidgets anxiously in his chair. The dialogue depicts an all-too-common exchange.

Candidates who experience the type of treatment described in this scene are not likely to ignore it; they will tell their friends, relatives, colleagues, and anyone else they know, to refuse to ever go to a job interview at that particular company.

Such an encounter can easily be avoided. For example, if the interviewer had required that candidates submit résumés in advance, had reviewed each résumé to determine suitability, and then had taken the time to set up phone-screens with potential candidates, this unfortunate scene would have been avoided. Phone-screens can prevent you and your interview team from wasting valuable time in useless interviews.

If you are not completely convinced that taking time to review résumés and to perform phone-screens makes sense, consider the following scenario: Imagine that you receive fifteen résumés. You review all fifteen, spending up to three minutes on each, and then decide to phone-screen six of the candidates. For those six, you spend fifteen minutes phone-screening each of four candidates, eliminating them from consideration, and thirty minutes phone-screening each of the remaining two candidates, both of whom you schedule for follow-up, in-person interviews. (You've now personally spent just under three hours reading résumés and phone-screening.) Next, you allocate three-quarters of an hour for each of your own interviews with both of the remaining candidates, and set up forty-five-minute time slots for each member of your five-person hiring team to interview each of the two candidates. The in-person part of the interview schedule comes to nine more hours (four and one-half hours per candidate). Adding the time you spent reading résumés (approximately one hour) plus the time you spent doing the preliminary six phone-screens (two hours) to the in-person interview schedule (nine hours) gives a total time expenditure of roughly twelve hours to pare fifteen applicants down to two fully qualified and professionally interesting candidates.

If you had chosen *not* to perform phone-screens first (saving about two hours of your own time), and you and your five team members each had interviewed every candidate in person who had passed your résumé filter (recall that this number was six in the scenario above), you'd each spend four and one-half hours for all the candidates in addition to the hour you spent winnowing down the original fifteen résumés (the number is derived by multiplying the six candidates times forty-five minutes per candidate-interview divided by sixty minutes), or a total of twenty-seven hours invested by you and the five interviewers after your original hour.

To figure out which approach makes maximum use of your five interviewers' time, I take your interview time of four and one-half hours out of the calculation and determine that the five members of your hiring team spent a total of twenty-two and one-half hours interviewing versus the seven and one-half hours the five-person team spent interviewing when you eliminated candidates by phone-screening them prior to setting up in-person interviews, yielding a difference of fifteen hours.

If I have the chance to save my employees fifteen hours of interview time by investing two hours of my own time to phone-screen, I'll take it!

By taking the time to phone-screen candidates, you will reduce the number of candidates, leaving only those you want to interview in person.[1] Phone-screens help you to verify that each candidate meets the minimum essential requirements. Further, they enable you to reduce in-

---

[1] The telephone interview as a screening mechanism is extensively treated in Martin Yate, *Hiring the Best: A Manager's Guide to Effective Interviewing* (Holbrook, Mass.: Adams Media Corporation, 1994), pp. 58ff., and in Del J. Still, *High Impact Hiring: How to Interview and Select Outstanding Employees* (Dana Point, Calif.: Management Development Systems, 1997), pp. 38ff.

person interviewing time and allow you to establish a rapport with candidates before meeting them face-to-face.[2]

Personally phone-screen the candidates whose résumés look promising. Don't take information passed to you by an internal or external recruiter as a substitute for information gathered during your own conversation with a candidate. When you talk directly with a candidate, the dialogue can provide you with valuable feedback on your résumé-filtering techniques. In addition, you may think of additional questions to ask during the phone-screen that you didn't originally consider during the job-analysis and job-description phases of the hiring process.

If your company is large or very formal and requires staff members from the Human Resources or Personnel Department to phone-screen candidates, take the time to train those people in the kinds of questions you want to use. Or, have HR personnel evaluate company and culture fit, rather than technical fit. Then, perform the technical-fit phone-screens yourself or with your technical team.

Facilitate a positive phone-screen environment.

When I contact a candidate to make an appointment for a phone-screen, I suggest several possible times, and then let the candidate pick whatever time is most convenient for him or her.[3] I call the candidate at the agreed-upon time, but at the start of the call, I verify that this is still a good time to talk. My reasoning is that while the candidate may have thought 6 P.M. would work well when we made the appointment at 10:00 that morning, it may no longer be convenient. If the baby is sick or the dog has gone crazy in the petunias and chased the cat up the neighbor's tree, the candidate will not be comfortable during the phone-screen. In such instances, I suggest we arrange a quieter, more relaxing time to talk. Once I've verified that the candidate is available to talk, I use my script to ask elimination-factor questions; open-ended, behavior-description questions; and wrap-up questions. (I briefly describe how to develop this script in the section below and add more detail later in this chapter.)

At the end of each phone-screen, I tell the candidate what the next step will be.[4] Specifically, if the candidate seems right for the job and I will

---

[2] The benefits derived from interviewing a candidate over the phone prior to meeting him or her face-to-face are presented in Lou Adler, *Hire with Your Head: Using Power Hiring to Build Great Companies,* 2nd ed. (Hoboken, N.J.: John Wiley & Sons, 2002), pp. 77ff.

[3] See Yate, loc. cit., for a sound discussion of the importance of mutual convenience in interview scheduling.

[4] Carol A. Hacker provides detailed guidance on how to end the telephone screen in *The Costs of Bad Hiring Decisions & How to Avoid Them* (Boca Raton, Fla.: CRC Press, 1999), pp. 60ff.

want to schedule an in-person interview, I say so—I also confirm whether the candidate is interested in proceeding further. If an in-person interview may be scheduled but is not a sure thing, I explain to the candidate that he or she may get a call to come in, but that it is not guaranteed. And if the candidate doesn't fit the job I have available and will not be interviewed further, I state that fact as well. At the end of every phone-screen, I make sure that each candidate understands where he or she stands in the hiring process, and I thank each for speaking with me.

Plan your phone-screen strategy and script.

I use a three-part approach in developing my phone-screens:

1. I ask for information about factors that could eliminate the candidate from consideration.
2. I ask for information about essential skills, preferences, and qualities.
3. I ask for specifics pertaining to personal or professional issues that could interfere with the interview process or with the candidate's ability to travel or perform overtime work required for the job.

These three elements are discussed more fully in the following sections.

Elimination factors

When you organize your phone-screen, plan to first ask about what could be *elimination factors,* such as a salary requirement, a benefits requirement, travel restrictions, availability to work overtime, and the like. The earlier you discover that a candidate's requirements don't match with yours, the better—you may want to ask about some of these factors even before you set up a time for the phone-screen.

In my experience, salary is often an elimination factor and so I tend to ask about it before taking care of any other piece of business, but my colleague Elisabeth Hendrickson takes the following approach:

"I never talk about money until I'm ready to make an offer. I can usually squeeze my management to make a higher offer for a really wonderful candidate. Further, I can usually offer really wonderful candidates something besides money—opportunity, growth, learning, and so on. I would hate to lose someone good just over salary when we won't know until we've

had a fair amount of time to get to know one another whether or not salary is a real issue."

If salary is not an elimination factor for you, then follow Elisabeth's lead and manage salary expectations at the end of the phone-screen when you've attracted the candidate to your company. I've rarely been in the position Elisabeth describes, where I could squeeze more money out of my management for a great candidate, and so I do make a point of discussing salary expectations. (The only times I've had salary flexibility have been for middle- to senior-level management candidates and equivalent technical staff.) I take a minute while I'm setting up a time for the phone-screen, and say to the candidate, "Annual salary for this position is in the forty- to sixty-thousand-dollar range. We generally hire people at the middle-to-low end of that range. Are we on the same page?" If the candidate says no, I ask how far apart we are. If we're too far apart, I decide whether it makes any sense to proceed. There's no point interviewing someone who wants a $120,000 salary, plus bonus, for a $70,000 position.

I handle matters in the same way if I'm working with a recruiter. In that case, I make sure the recruiter knows the salary level for the open position, and has explained that level to the candidate.

If you and the candidate are far apart on salary, try to figure out why:

- *Is the candidate more senior than you need for the position?* (If so, decide how flexible you are with salary now, before you've wasted your time and the candidate's time.)
- *Is the candidate more experienced than his or her résumé led you to believe?* (If so, decide how much time you'll spend on the phone-screen before you decide whether to stop the interview process here or continue.)
- *Is the salary you are offering far below the industry or regional standard?* (If you've phone-screened a number of candidates and they all want more money, it's time for a salary survey to verify that the talent you're looking for matches the money you're willing to pay.)

If your company doesn't set salary ranges according to job categories, maybe salary is not truly an elimination factor, but you probably will have other elimination factors aside from salary. A person's willingness to travel and his or her availability to work (that is, to begin the job on the date you've set, or to work overtime) are the two most common factors, but if you have other concerns, ask about them in the phone-screen.

You may have technical-skill elimination factors also. Although I recommend that you cover skills, preferences, and qualities as the middle portion of the phone-screen, if you absolutely require, say, C++ and UNIX experience, make sure you ask early in the phone-screen whether the candidate has this experience.

Essential skills, preferences, and qualities

For all candidates who have made it through to the phone-screen without running up against any elimination factors, I next ask about skills, preferences, and qualities. I choose questions that will help me determine whether to bring the candidate in to interview in person. I try to obtain a whole picture of the candidate, and so I ask a mix of questions. For example, if I want to know about a designer's essential functional skills and how he or she applies those skills, I might say, "Tell me about a recent design problem you encountered. Tell me what was challenging about it and how you knew the design was complete."

If I want to know about a tester's functional skills and his or her understanding of the test tools in a particular environment, I ask the following kinds of questions: "When did you choose to use this test tool?" "What kinds of testing did you perform with this tool?" "What did you learn from that testing experience?"

To learn about a technical support representative's knowledge of the industry, I might say, "Tell me about the kinds of customers you have, and what industries they are from." I listen to the answer, and then draw out more information by asking, "Do you have to perform different work for different customers?"

If you are asking about essential qualities and preferences rather than about functional skills, your questions will be a bit different from the ones I've just suggested. Suppose that you are interviewing for a technical support representative who will have specific on-call hours. You can ask, "How have you handled on-call work before?"

Use open-ended questions to understand the candidate's experience in the essential qualities, preferences, and skills. If I have a junior-level candidate who has little work experience, or a candidate whose answers aren't particularly focused, I change my questions to be more specific. For example, if I'm interviewing a junior-level developer and I want to know about his or her debugging skills, I might say, "Tell me about your current project." Then I might ask a series of questions: "What's challenging about that project?" "Have you run into any challenging bugs?" "How did you determine where the problem was?" "How long did it take you to fix the problem?" I wait for answers between questions, of course, but I rely on the questions to lead the candidate rather than leave him or her struggling to guess what I am looking for.

If I want to know how a project manager handles risk, I might ask, "How did you identify risks on your last project?" If the candidate says, "Oh, I made a risk table," or "Oh, I held weekly meetings with the team during which we discussed areas of potential risk," I ask follow-up questions, which are designed to expand upon the information the candidate has provided: "What were the risks?" "Did you have trouble identifying information to enter into the table?" "What did you do with the risk table?" "Were you able to use the table to discuss risks with your management as well as the project team?" Use the job-analysis table and the job description to select *which qualities, preferences, and skills* you want to ask about in the phone-screen. Then, decide *how* you will phrase the questions.

Personal and professional issues

Once you've learned what a candidate's essential skills, preferences, and qualities are, and you're satisfied they match your needs, ask additional questions to determine whether the candidate has other requirements that your job will not satisfy. These wrap-up questions should not cover the same ground as was covered by the elimination questions you asked at the time you scheduled the phone-screen, but should delve into topics not previously addressed. The information you get should help you decide whether to schedule an in-person interview.

- *"Why are you leaving your current position?"*

People leave jobs for all kinds of reasons, but I tend to be wary of the candidate who wants to leave a job because he or she is unhappy with the current work environment. I worry that such candidates are not evaluating my open position on its merits, but on how quickly taking it will get them out of their current job. During my in-person interview, I assess whether the candidate appears genuinely interested in the work I have for him or her. I do not want to hire people who intend to continue the job search after they've started working for me—that's an expensive hiring mistake. If you discover that a candidate is unhappy in his or her current environment, probe further. If the candidate is looking for a greater challenge or believes that he or she is underpaid compared to the market, don't be too worried. Do some research to verify whether the reasons hold up. Of greater concern is the candidate who tells you the culture is toxic or the organization is chaotic or the commute is unbearable. Probe further to determine what made things toxic or chaotic or unbearable, and mentally check to see whether your company has any of the same problems. It is not a crime for a candidate to be unhappy with his or her current job or to want to better his or her lot, but a mature, professional person will be circumspect about harping on the problems and will focus on finding an environment in which to be productive. If a candidate excessively criticizes management policies or colleagues at his or her current company, think about whether the problems are the candidate's problems or the environment's problems.

- *"What are your current and asking salaries?"*

Because it is possible that a candidate has not been paid a salary commensurate with his or her responsibilities and skills and may be looking for a substantial increase, it is essential that you know the facts. If what the candidate is currently earning and what he or she hopes to make are significantly different, I ask directed questions about the candidate's current responsibilities to make sure the increase is justified.

- *"What is your base salary?"*

In order to encourage the candidate to tell me the absolute truth about his or her current salary, I ask for the base salary, and I record bonuses and other perks separately from base salary. (Note how this question differs from the previously asked salary-range question.)

- *"Have you had any interviews that you expect will result in a viable offer?"*

Finding out early on whether you have competition for a candidate helps you to assess the candidate's value to you, but you should not let this knowledge force you to make an early decision. I learned a valuable lesson early in my career when a candidate told me he expected several offers, and named the companies he thought were interested in hiring him. I knew the hiring manager at one of the companies he'd named and believed I could speak frankly with her as a fellow professional, and so I asked her what she particularly liked about this candidate. She looked surprised and explained that she wasn't going to make him an offer. It turned out that the candidate was trying to force me into an early offer.

- *"What is your availability next week for a round of in-person interviews, and when could you start, if hired?*

I ask this two-part question because I want to confirm that the candidate will be available to be interviewed within a reasonable period of time, and because I need to know when the candidate could start work. Unless there are extenuating circumstances, I expect people to be available for in-person interviews during the calendar week following a phone-screen. I also expect that people will want to give a minimum of two-weeks' notice to their current employer, and that some may also want to take a week or two off between jobs. However, I don't expect people to tell me they are not available to start for three months so they can finish their current project.

Select phone-screen questions to elicit job-performance details.

I ask behavior-description questions when I want to learn more about events that happened to the candidate, and when I want to know about his or her reactions to those events. As I noted earlier, if I have a junior-level candidate, or a candidate whose answers aren't particularly focused, I modify my questions to help the candidate focus more specifically on information I need to know.

I may ask a meta-question if I'm not sure I've received a complete answer from a candidate. For example, I may ask, "Is there more I should know about this?" Or, if I have established a good rapport with the candidate, I might say, "It sounds as if there's more to that story. Is there?"

You may even be able to hold a kind of mini-audition during the phone-screen. Although I have not found a way to hold anything that amounts to a true audition, I do ask questions to try to draw out a picture of how the candidate would perform in certain types of jobs. If you have developed an audition that you use when you and the candidate are not in the same room, test it with someone already successfully performing the same role in your organization. Table 8-1 contains some questions that can be posed during a phone-screen to elicit information on how candidates for different positions would handle a specific challenge.

Job Title	Opening Question	Follow-up Questions
Developer	How would you create a random-number generator?	When was the last time you worked on something like this? What was it? What happened?
Tester	How would you approach testing a random-number generator?	When was the last time you tested something like this? How did you test it?
Technical Support Representative	How would you go about documenting the problem of a customer calling up to complain that the random-number generator we have in our product isn't random?	When was the last time you ran across a problem like this? What was it? What did you do?
Writer	How would you document a random-number generator?	When was the last time you documented something like this? What was it? What did you do?
Project Manager	How would you rescue a project in trouble?	When was the last time you rescued a project in trouble? What kind of project was it? What did you do?

Table 8-1: Performance-Assessment Questions to Ask During the Phone-Screen.

Don't stop with your opening question—follow it up with open-ended, behavior-description questions. Questions that assume some hypothetical situation can be used, but be careful to ensure that the candidate understands the question and that you understand the answer.

Avoid irrelevant questions on the phone. Irrelevant questions don't sell the candidate on your company. Nor do they provide you with information about the candidate's qualities, preferences, or skills.

Use written phone-screen scripts to keep track of what candidates say.

A written phone-screen script can help you ask each candidate the same questions you ask all other candidates, and provides a place to record notes about each candidate's answers. There's no law that I know of that requires me to ask each candidate the same set of questions, but I like to be able to compare answers. I find being able to compare answers especially valuable when I'm interviewing numerous candidates, when I'm interviewing people for a job for which I've not previously screened candidates, or when I'm trying to decide which candidates to schedule for an in-person interview.

Before I set up any in-person interviews, I use the script to confirm that I have asked *each candidate* questions that could rule him or her out of contention. I also use the script as a prompt for asking the open-ended, behavior-description questions that will help me learn more about the candidate's work attitudes and habits. An additional benefit of the written script is that, often, as I jot down what the candidate has said, he or she will fill in the pauses, frequently providing me with more information.

In my experience, technical candidates respond well to questions that help them tell me their stories. When I ask for information, saying, "Tell me about how you helped that project succeed," or "Give me an example of how you solved a lingering customer problem," the candidate knows that I am serious about the position and want whomever I hire to succeed in it. My phone-screen questions not only reveal what's important to me, they also can indicate what it might be like to work in my company and even with me.

Develop a thirty- to forty-five-minute phone-screen script.

Once you've decided to use a phone-screen, make it long enough so you can properly assess the candidate's suitability for the job. I find that a phone-screen that takes less than thirty minutes is not long enough for me to assess *a qualified and suitable candidate* properly, but some authorities do suggest only a twenty-to-thirty minute slot.[5] A length of thirty minutes *is* long enough, however, for me to determine when a candidate is not well-suited to the job I have open—a topic I discuss in greater detail later in this chapter. For suitable candidates, I try to end the phone-screen at about the forty-five-minute point. If I'm on the phone any longer than forty-five minutes, I'm probably asking questions better asked in an in-person interview. I end the phone-screen when any one of the following happens:

- I've used up forty-five minutes.
- I want to bring the candidate in for an in-person interview.
- I want to ask questions better answered in an audition.

If you don't need the time, feel free to end the phone-screen before the thirty-minute mark. One reason to bring up elimination factors early in the phone-screen is to use them to weed out candidates who won't fit. Even though I allocate thirty minutes for phone-screens, I do end them early when I realize the candidate won't fit my job.

I use the same phone-screen script when I am interviewing contractors as I use to screen potential employees. If I'm going to take the trouble to interview a contractor, I want to be sure that the contractor is a good fit. Some hiring managers use an abbreviated phone-screen for contractors, but I have not had good results doing so.

Troubleshoot your phone-screens.

If you've tried phone-screens, but haven't always found them helpful, review the questions you ask. Check that you first ask elimination-factor questions and then ask open-ended, behavior-description questions, so

---

[5] See Hacker, op. cit., pp. 60ff.

that you can learn enough about the candidate within forty-five minutes to know whether you want to proceed.

One technique some screeners use is the "what-kind-of-a-person-is-this" phone-screen to confirm that the candidate is not a rotten apple. If you only want to phone-screen or interview people who are courteous, ask an administrative assistant to place an initial phone call setting up your phone-screen. Your assistant can assess how courteous the candidate is, thereby testing one of your elimination factors.[6]

When you bring someone in for a face-to-face interview, and discover that the person is not a good fit, the discovery gives you feedback about the quality of your phone-screen. When this happens, think about what questions you might ask in future phone-screens to eliminate such candidates from consideration before the face-to-face interview.

If you're asking good questions, and you're still unsure about the candidate, take your uncertainty as information about the person. Ask yourself questions such as, "Is this candidate answering questions so that I can't tell if I've heard the information I need?" "Is that acceptable behavior in someone I would hire for this job?" If you can answer no to the first question and yes to the second, proceed with setting up an in-person interview. If not, decline to interview the candidate.

Some years ago, I finished a phone-screen feeling that the candidate was not a good match for the position, but I couldn't put my finger on why I felt that way. After I'd phone-screened more people, I realized that the first applicant had only half-answered my questions. Because I'd written down his answers, I was able to go back to compare his answers, as noted on my phone-screen script, with other candidates' answers. It was then that his lack of specificity became apparent—another reason for asking the same questions of each candidate.

End the phone-screen gracefully and when *you* want to end it.

You can end a phone-screen at any time. If a candidate doesn't make it past your elimination factors, you can confidently end the phone-screen. When I was hiring testers who needed to know both how to use particular kinds of data structures and how to develop data-structure tests, I asked all candidates for the job to tell me about their experience using linked lists. If they didn't know what a linked list is or had no experience using one, I felt confident that they were not right for the open job.

---

[6] A startling description of how some corporations use phone-screeners to weed out candidates at various stages in the hiring process appears in Liz Ryan, "HR Interviews Can Be Senior Execs' Waterloo," posted at *The Wall Street Journal*'s Executive Career Website (www.career.journal.com/columnists/perspective/2003/006-fmp.html).

Candidates may make it past elimination-factor screening, but, later in the phone-screen, they may reveal their support of a practice or behavior that doesn't fit with the job. If so, investigate further by asking several more behavior-description questions. If the ensuing answers are not job-appropriate, I usually end the phone-screen there. The longer your phone-screen lasts, the more likely the person will think that he or she is a suitable candidate, and the harder it will be to end the process.

If you're new to hiring, you may have trouble ending the phone-screen early. Don't despair—there are reasons you can give for ending the phone-screen early that won't make the candidate angry or believe that you have misled him or her. Two possible explanations you can give for terminating the phone-screen follow:

- You can explain that you appreciate the time the candidate has given you, and that you don't need to take up any more of it at the present.
- You can explain that experience with a specific environment/language/tool is a requisite for the job and that the candidate's experience does not match.

There are numerous other logical and justifiable reasons to give, but if you're still not sure what to say to end the phone-screen, try one of the three closings suggested below:

- *Yes-we-want-you* closing remarks explain to the candidate that you're interested, and that you want to continue the interview process: "You seem like you might be a good fit for this position. Harriet in Human Resources (or Samantha in Systems Analysis or Terri in Test Engineering, and so on, as applicable to the open position) will call you tomorrow to arrange an in-person interview. What number can she use to contact you?"
- *Maybe-we-want-you* closing remarks should indicate to the candidate that he or she is not at the top of your list, but that you or someone from your company may be back in contact to set up an in-person interview: "You have some of the experience we need, but not all. If we want you to come in for an interview, Mr. Personnel will contact you, probably within the next two weeks. What number should he use?"
- *No-we-don't-want-you* closing remarks indicate to the candidate that you are looking for a different mix of qualifications or for specific experience he or she doesn't have: "I appreciate your taking the time to talk to me, but your mix of skills

and experience isn't quite what we're looking for. I wish you every success in your continued job search."

Remember, you don't need to give the candidate a long and involved reason unless you want to explain your decision. You simply need to thank the person for his or her time and then tell the candidate where he or she stands with you. Some candidates realize from the questions you ask during the phone-screen that they don't meet your needs, and will amicably end the conversation. However, sometimes people expect to be interviewed in person, and become quite aggressive if they suspect they will not have that chance. Here's the conversation that transpired when I tried to close one such phone-screen:

> Me: "I'm sorry, but you don't meet our requirements for this position. Good luck with your job search."
>
> Candidate: "Well, exactly how don't I meet the requirements?"
>
> Me: "You don't have the problem-solving and communications skills we're looking for."
>
> Candidate: "What do you mean I don't have those skills? I told you about that problem I solved four years ago. And furthermore, I communicate all the time."
>
> Me: "That problem-solving experience happened four years ago, and although I asked you to tell me about more recent, relevant experience, I didn't hear anything that jibed with what we're looking for. I'm also looking for a person whose communication skills fit with a group that has worked together for some time and that dislikes confrontation. I don't think you'd fit well with that group."
>
> Candidate: "Oh, but . . ."

What I felt like saying—but didn't because saying more usually makes confrontational people even more aggressive—was, "And, listen to yourself! You're confronting *me* in what should be a friendly, preliminary phone-screen interview! I'm looking for a person with specific, recent experience to fit into a group that actively dislikes confrontation. You don't fit that group. You'd hate it, and so would each and every current team member hate having you in it."

You have several choices when a candidate doesn't want to let you terminate the interview process at the end of the phone-screen:

- You can simply say "thank you and goodbye" and hang up.

- You can explain as I did.
- You can explain (or not), and suggest alternative employers whose open positions might better suit the candidate's qualifications or personality.

Remember, *you* own the phone-screen. If you're uncomfortable rejecting candidates, practice saying your version of "Thank you, but no, you don't fit," so that you'll be ready to provide the negative response at the end of the phone-screen.

Be aware that your biases can prevent you from making the right decision at the end of the discussion. Sometimes, I want to avoid what the phone-screen is telling me, because there's something about the person that, despite negatives or disqualifiers detected during the phone-screen, still seems to indicate a viable candidate, and I want to bring the person in to see whether my instincts are right.

If you catch yourself trying to make the candidate's phone-screen answers "right," ask yourself, "Am I having trouble deciding because of something about the candidate that has nothing to do with the questions, or is there an answer in the phone-screen that's not specific enough?" If you can't decide, buy yourself some time and give the maybe-we-want-you closing remarks.

Consider when to use a second phone-screen.

Consider conducting a second phone-screen if you believe bringing in a candidate for an in-person interview would be too costly, either in terms of interview time, travel expense, or the like. If you do schedule a second phone-screen, ask substantive technical and behavioral questions, and create an audition-like setting.

You don't have to limit yourself to the phone; video-conferences may be an alternative, especially if you want to see how the candidate conducts himself or herself before you proceed with scheduling an in-person interview. Keep in mind, however, that neither phone-screen nor video-conference interviews are a smart alternative to face-to-face interviewing if you are going to make a firm offer to a candidate. Imagine the following happening: You hire a candidate who has passed phone-screen and video-conference interviews with flying colors, and then discover two unsettling truths when meeting your new hire in person for the first time: Your new hire is a highly qualified, technical wizard who (1) smells like unwashed socks and (2) has marginal social skills. Although body odor or substandard social skills may not be enough to cause you not to hire a candidate, you may want to know about those details in advance.

Although I don't recommend that video-conferencing be used as a replacement for an in-person interview, video-conferencing can provide visual and aural information about a person.

Case Study: Walker Software

Dirk, the Development Manager at Walker Software, prepared the following phone-screen script to use while conducting a job search for Software Developers with specific telephone-industry experience using C++ and Unix operating systems. I've added my observations in italics:

Software Developer   Yes _____ Maybe _____ No _____
Candidate name:_____ Phone #:_____ Date:_____

1. "Let's make sure we're on the same page with respect to salary. This position is a developer position, with a salary range of $40,000 to $70,000. We don't normally bring people in higher than at the mid-point, $55,000. Are we on the same page?" *(If Dirk were working with a recruiter, he would have already discussed salary range with the recruiter, but when he contacts a candidate directly, he asks the salary elimination-factor question.)*

2. "How many years of C++ and UNIX experience do you have?" "How many years have you had handling UNIX system calls?" "How many years of UNIX shell-scripting have you had?" *(Dirk starts with close-ended questions to establish the facts and to eliminate candidates without enough experience.)*

3. "Tell me about your work in the telephony industry." *(Dirk has made this request open-ended to see what the candidate says. If the candidate doesn't say anything specific, Dirk can then ask about performance and reliability, such as, "When was the last time you worked on a performance or reliability problem? Tell me about the problem and how you solved it.")*

4. "Have you had a role in defining and using data structures in your current project?" *(Dirk wants to know what level of responsibility the developer has had, and whether the developer has done some design and/or debugging.)*

Once Dirk has asked elimination-factor questions and has verified that the candidate has the essential technical skills, he next will ask a set of questions to address non-technical essential skills, preferences, and qualities, as follows:

5. "What can you tell me about the team you're on now?" *(Dirk is looking for someone with the ability to work well on a team. He should follow up with questions such as, "What's your role on the team?" "Do you ever pair up and do pair programming?" "What happened?" "Has your team evolved over time?")*

6. "How did you decide on the design on your current project?" *(Dirk's question may be difficult to answer in a phone-screen, where the candidate probably cannot demonstrate his or her ability to draw pictures, but it's worth trying to ask the question in some way. Alternative questions could be: "Have you ever evaluated designs, and if so, what did you use for evaluation criteria?" "How do you know a design is good?" "When do you stop designing?")*

7. "Have you had to change the focus of your work, and if so, when was that, and what happened?" *(Dirk is looking for evidence of a candidate's degree of adaptability.)*

At this point in the phone-screen, Dirk presumably knows about the candidate's essential preferences, skills, and qualities. If he's interested in the candidate, he'll wrap up with a final group of questions.

8. "What are your current and asking salaries?" *(Even if Dirk is working with a recruiter or has asked the initial salary question, this is a good time to re-check the facts.)*

9. "Have you had any recent interviews that you expect will result in a viable offer and, if so, what salary range is likely to be offered?" *(Dirk wants to verify that the candidate is not on the verge of accepting another job offer and that he or she is seriously interested in the job Dirk has to offer.)*

10. "What's your availability for an in-person interview?" *(Dirk doesn't have time to waste, and he wants to make sure the candidate is interested in moving quickly.)*

11. "If you were offered this job, when could you start?" *(This question further probes the candidate's commitment.)*

12. "Why are you leaving your current position?" *(For a candidate who is not currently employed, Dirk would ask this in the past tense: "Why did you leave your previous position?" Dirk wants to know that the candidate is looking for something, not just moving away from his current position.)*

POINTS TO REMEMBER

- Develop a phone-screen script for each open position.
- Ask elimination-factor questions and determine essential skills, qualities, and preferences during the phone-screen to help decide whether an in-person interview is merited.
- Ask the same questions of all applicants for a given position.
- End the phone-screen if the candidate doesn't pass your elimination-factors questions.
- Conduct phone-screens that are as short as five to fifteen minutes as you search for candidates worthy of the full thirty- to forty-five-minute phone-screen or of an in-person interview.
- Decide whether to invite the candidate in for an interview, whether to put his or her résumé in the Maybe pile, or whether to reject the candidate entirely. Tell each candidate what he or she can expect at the end of the phone-screen.

Planning and Conducting the In-Person Interview

Interviewer: "Have a seat on this step while I take a few seconds to look at your résumé."

Job Applicant: "Are we meeting in this stairwell for the interview?"

Interviewer: "Well, yes. No one booked a room for us to talk in but we'll have lots of privacy here because no one ever takes the stairs."

The preceding dialogue indicates how ridiculous an interviewer's lack of planning prior to meeting with a candidate could seem, but it depicts a realistic exchange extracted from a senior-level developer's True Story (printed in full on the following page). Don't let this happen to you when it's time to plan the interview. Interviews should serve two distinct purposes: First, they should help you discover how the candidate has worked in the past so you can determine whether he or she is suitable to fill your open position. Second, they should help you sell the candidate on your company. Even if you don't hire the candidate, he or she may refer others to you, so it is important that you make the experience as positive, seamless, and professional as possible.

There are steps you can take to assure that the process runs smoothly, and that it is positive for the interviewers as well as the candidates. You will need to choose an interview team; plan who will ask the candidate which questions about skills, preferences, and qualities; and create and distribute an interview package so that all interviewers know what to expect. If you take the time to create an interviewing environment that reflects how you work—that is, one that is true to the corporate culture—you will benefit from having an interview process that works for you.

A True Story

"I walked up to the receptionist and asked for Joe Smith, the VP of Engineering. I'd arrived about ten minutes early for my 9 A.M. interview, but the receptionist rang Joe's extension. No answer. She then paged him. Still no answer. She told me to take a seat until she could locate Joe, so I sat.

"At about 9:10, a harried woman flew into the reception area and asked me for my name. I told her and she immediately took me by the arm. 'I'm Amanda. I'm sorry to be so late,' she apologized, 'but I just heard you were coming in for an interview. Let's go.'

"We walked through the fire doors and into the stairwell, where she sat down on one side of the bottom step. 'Now, let's see. Have a seat on this step while I take a few seconds to look at your résumé.'

"As she settled into reading, I interrupted. 'Are we meeting in this stairwell for the interview?'

"'Well, yes,' she said.

"'Why are we in a stairwell?'

"'Oh, for privacy.' Amanda patted the step for me to sit down next to her. 'No one booked a room for us to talk in but we'll have lots of privacy here because no one ever takes the stairs.'

"So there I was, in a stairwell, talking about my work experience and career goals to a woman I didn't know. And, of course, during the next five minutes, three people came running down the stairs on their way into the reception area, interrupting our conversation, and making me lose my concentration. I wanted to walk out, but I really needed a job. I had to ask myself, 'Do I want to take a job here?'"

—Discouraged job-seeker

Choose an interview team.

My objective in choosing an interview team is to gain insight into how the person potentially would work with the specific mix of people in the specific environment. I begin by creating an interview team made up of the *people who will work with the candidate if he or she is hired.* If the candidate could work with more than eight people if hired, I choose from those people—up to a maximum of eight—to serve as members of my interview team, but I do not use all of them for the first-round interview.

In fact, I prefer to limit a first-round, in-person interview team to four or five people, plus the hiring manager. My reason for doing this is to insure that, combined, we spend no more than half a day assessing a candidate (recall from the previous chapter that I allocate forty-five minutes per interviewer per candidate plus my own forty-five-minute session if I am the hiring manager). If we decide, for whatever reason, not to proceed with the candidate, we haven't wasted much of any one person's time—including the candidate's.

For the second round of interviews, I again use people from the department, project, or geographic area where the candidate will be assigned. For example, if whoever fills the job opening will work with more than eight people, I choose another four or five representatives from the people with whom the candidate will work. If the candidate will be working with many people, such as someone hired as a project manager to manage an 80-person project, I may select more management representatives so that project staff members, peers, *and managers* all have a chance to interview the candidate. My goal is to select up to a total of ten other people to conduct the first and second rounds of interviews.

If the candidate will work with people spread across the organization, I choose representatives—up to ten, of course—from a variety of groups and departments to participate in the interview. For example, if I were going to interview a candidate for the position of a senior-level developer, I would choose four or five technical peers, one project manager, one writer, one or two senior-level testers, plus myself (assuming that I am the hiring manager), to form the interview team. This selection further assumes, of course, that I have a big enough pool of people already performing these roles from which to draw this eight-to-ten-person interview team. If the developer will work more with requirements than with documentation, I would substitute a systems engineer, a business analyst, or a product manager in place of the writer—that is, someone who defines requirements. For the first-round, in-person interview, I would request three of the peers and one of the specialists to participate in the interviews. And so on.

If you're in a highly competitive job market and plan on making a same-day offer, you'll need to assign more interview participants to the first round than the four or five I generally recommend. This is because more interviewers will be needed to gather enough information about the candidate in a compressed time frame. Conversely, if the candidate makes it to the first-round interview session or somehow slips through to the second round, but is obviously unsuited for the position, plan in advance with the full team how the session can be interrupted mid-session and canceled—and plan by whom this action can be taken.

Choose whether and how to use your Human Resources representative on the interview team. If the representative is not capable of participating in an interview designed to assess the candidate's preferences, qualities, and skills, then I invite the HR rep only to the second round.

When planning the makeup of my interview team, I look for people who already know how to interview. If I'm working with a department or in an organization where people have not yet learned how to ask behavior-description questions, I set aside a half-day before the first interview to teach potential interviewers how to ask questions, and follow up with those interviewers just before the candidate arrives.

I prefer to keep the selection of interviewers consistent when interviewing for the same position, so that I have a consistent set of questions, biases, and approaches to candidates. You don't want to be in the position of bringing an interviewer in at the last minute and having the interviewer say, "Well, I think this person is a good candidate, but of course, I didn't get to interview the others, so I can't compare candidates."

Prepare the interview team.

Your interview team needs to be prepared before the interview, and preparation takes time. If you repeatedly use the same people for interviewing, be considerate of the time required of those people both to get prepared and to interview. Time taken away from project work is costly, but *not* taking the time needed for preparation is worse, because the interviewer wastes the interview time by not being adequately informed.

Here's how I approach the problem: First, I send an e-mail requesting that my interviewers set aside specific times in their calendars for interviewing. I use calendar software to help me schedule and follow up with the interview team, but sometimes, the HR rep or an interview coordinator can help schedule the interview team if these people are available to you.

Next, I talk to interview-team members and verify that all members of the team share the same goals for hiring for this job. I hold a quick team meeting to focus people on the similarities and differences between this job's description and previous jobs they've filled. By making sure the interviewers understand the job description, I can feel fairly confident that people won't veto candidates because they don't understand the role we want filled. If an interviewer has never seen a job analysis before, or hasn't considered candidates except on the basis of technical experience, the job description may not mean much. I also always try to make sure everyone understands just why we're hiring another person. I ask whether anyone has anything to fear from this potential hire, and I deal with that fear now, before people start interviewing candidates. I have learned that if I don't clear the air of any fears about job security, I won't be able to hire anyone because I won't be able to get honest appraisals of candidates.

In addition to convening an interview-team meeting, I frequently hold one-on-one meetings with individual members of my team to give people a chance to raise any of their concerns about this hire before we start interviewing.

After I send e-mail to interview-team members to request time in their schedules, I distribute a hard-copy interview package to each interviewer. The package contains the candidate's résumé, the job description, and the schedule of who will interview each person when, as well as a listing of who will focus on which question areas. I don't send this important documentation as an electronic file because I want people on my interview team to spend as little time as possible on administrative tasks—such as printing the interview package. I want people to use their time preparing for and participating in the interview, and I want them to still feel they have some time left to follow up with me after the interview. Another reason why I don't send details electronically is that too often an interviewer will forget to print the package ahead of time—leaving too much to memory instead of having words on paper.

Part of the hiring manager's job in preparing interviewers is to help them decide what to do if someone on the interview team hits a brick wall with regards to the candidate. For example, sometimes an interviewer may feel that he or she could not work with the candidate for some reason or another. In the preparatory session, I discuss what options exist in such a case. There are several choices: The interviewer can be asked to persevere; the interviewer can end the interview and walk the candidate out; or the interviewer can walk the candidate to me, so that I can decide what to do.

In the preparatory session, I instruct interviewers always to hold visceral or subjective reactions at bay when they ask a set of objective questions written for just such situations. However, if an interviewer feels he or she has reached the halfway point in the interview, has asked the objective questions, and still thinks there's no way the particular candidate could succeed in the job, my preference is that the interviewer (1) tell the candidate there are no further questions, (2) ask the candidate to wait in the interview room, and (3) come directly to my office. I take a few minutes with the interviewer to discuss the problem; I may then check with other interviewers to see if anyone else has had the same reaction. In most instances in which an interviewer has felt strongly that the candidate would not work out, I've found the assessment of the candidate to be accurate. Whenever this is the case, I ask some wrap-up questions, thank the candidate for coming in, and then I walk the candidate out.

Decide how much time to spend in each interview.

As stated previously, I prefer forty-five-minute interview slots. That way, I have enough time to ask enough questions in depth—and hear the answers. If you're not sure how much time you need, look back at the last few positions you filled. Ask yourself questions such as the following:

- Did I want to ask more questions than the time allotted?
- Did I need more time to ask questions about my assigned areas?
- Did I need more time to draw out information about the candidate's technical expertise or personal preferences?
- Did we have to bring the candidate in for more than eight hours of interviews or for more than two interview rounds?
- Did we miss cultural-fit issues?

If your answer to any of these questions is yes, look at who's interviewing the candidates and how much time each interviewer spends with each candidate. If you're not spending enough time during an interview, you'll want to ask more questions and may need to bring the candidate back several times to learn more information. If you missed cultural-fit issues, then it may mean you do not have skilled interviewers on the team, or they might not be spending enough time with each candidate. Maybe you need to add an interview slot for an audition, to verify the candidate's technical expertise. All the interview time in the world won't help if the interviewer doesn't know how to ask questions to obtain information from a candidate. Think about how much time each person will spend interviewing each candidate. No matter what you choose, decide in advance how long each interview slot will be. Each interviewer does not have to take the same amount of time, but you should establish parameters for consistency. And whatever time slots interviewers choose, leave enough time after every couple of slots for candidates and interviewers to refill water glasses or take bathroom breaks.

Auditions, described in greater detail in Chapter 7, "Developing Interview Questions and Techniques," may take longer than forty-five minutes. When you're planning the interview, make sure you know who will conduct the audition and that the audition time slot is long enough for a candidate to demonstrate both ability and knowledge.

Plan who will ask which questions.

After I've selected people for an interview team, verified that they know how to ask appropriate interview questions, and made sure that we've allowed sufficient time for each interview, I assign question areas to each member of the team.

You may assign the question areas yourself or let members of the interview team decide what questions they'd like to ask. Whichever you choose, deciding beforehand who will ask what helps you ensure that members of the interview team do not repeat questions. The more question variety, the better the overall picture of the candidate you'll get.

Take your job analysis or job description and meet with the interviewers to plan who will ask which questions. When you're planning, decide how many areas each person will focus on. I recommend that people explore no more than two general areas. If you're trying to cover too many areas in an interview, you might settle for the first obvious answer instead of asking more follow-up questions. On the other hand, if you don't have enough to ask about, you'll feel lost in the interview.

Table 9-1 provides a scheduling matrix for interviewing a technical manager and indicates the areas each of five interviewers needs to explore. Note that although some interviewers have x-marks in more than two topic areas, there is some content overlap and the areas do not exceed "general" areas. Note also that a brief break has been built in between Interviewer Three's time slot and Interviewer Four's start time.

When you create your interview matrix, try to assign interviewers so that you're asking questions about the most important areas earlier in the interview. That way, if an interviewer decides the candidate is not right, that interviewer can end the interview.

Each interviewer has a unique set of two or three areas about which to ask questions. Before the in-person interview, Interviewer One also phone-screened the candidate in order to ask the elimination-factor questions and some questions about the essential qualities, skills, and preferences. Interviewer Two normally prefers thirty-minute time slots, but the grid shows us that he is willing to spend forty-five minutes if he can ask more questions.

	Interviewer One	Interviewer Two	Interviewer Three	Interviewer Four	Interviewer Five	Everyone
Time	8:00-8:45	8:45-9:30	9:30-10:15	10:20-11:05	11:05-11:50	11:50-12:05
Location	Conference Room A	Interviewer Two's Office	Conference Room A	Conference Room A	Interviewer Five's Office	Conference Room A
Question Areas						Meet to evaluate the candidate
Planning Skills	x				x	
Project-management Skills		x	x			
General Problem-solving Skills	x				x	
Decision-making Skills		x	x			
Multitasking Skills		x		x		
Technical Process and Methodologies	x			x		

Table 9-1: Sample Interview Matrix for a Technical Manager's Position.

I used to provide my interview-team members with behavior-description and other types of questions, but I discontinued that practice when I realized that interview-team members read résumés more closely when they need to decide how to ask their own questions of each candidate. Now I encourage members of the interview team to develop their interview questions to reflect the context of the candidate's résumé.

When working with people who are inexperienced interviewers, I use the preparation meeting to perform a dry run of at least one interview question from each interviewer. I ask people to move into groups of two or three, with one person playing the role of the interviewer; another, the candidate; and the third, if available, the observer. To role-play with three in a group, the interviewer asks the candidate a question, the candidate responds, and the observer helps the other two understand what worked with the question and what didn't. In pairs of candidate and interviewer, people must debrief themselves, but the process is still effective. At the end of each question-and-analysis enactment, I ask people to change roles and practice another question.

I particularly want members of my interview team to develop their own cultural-fit questions, addressing qualities, preferences, and non-technical skills. If I, as the hiring manager, ask a question about a candidate's ability to manage multiple projects, I may hear answers that differ dramatically from what a technical person asking the same question will hear. For example, if a member of the interview team questions the candidate about working for more than one project manager at a time, or about serving on a project whose vision the candidate doesn't share, that interviewer may hear a different answer than would be offered to a potential manager. When interview-team members choose their questions, the questions may be seen as more genuine, encouraging candidates to answer completely and from the heart.

The one event I don't usually ask interviewers to prepare their own questions for is the technical audition. As an interview *team*, we develop the audition questions together, and then choose one interviewer to conduct the audition.

Sometimes, especially during preparation for a second-round interviewing session, you may assign several interviewers to ask the candidate questions about a subset of skills, preferences, and qualities. In such a case, have interviewers plan how their interview questions will be different. Interviewers can ask questions about different parts of the candidate's résumé, or can take different approaches to questions. If you haven't already chosen audition questions, this is a good time to do so.

Choose an appropriate interview environment.

I've heard wild stories about interviewing environments that would make a candidate's hair stand on edge:

- a well-traveled stairwell
- a busy lobby
- a cafeteria at lunch hour
- a train station during morning rush hour

While I can imagine that it is possible to conduct an interview in the midst of chaos, I can't imagine things going well. But I once was told of an experience that does seem unimaginable—the concluding minutes of the interview occurred while the interviewer changed his baby's diaper! An interview environment should be chosen to help create an atmosphere in which people can speak in a professionally amicable fashion.

For the interview, find a private area where you will not be interrupted, such as a conference room, a private office, or a secluded area of the cafeteria during off-hours. I prefer to interview in a room with a door, so that I can close the door and have a private conversation. An interviewer should try to create an atmosphere in which the candidate can think and respond, without feeling pressured by external factors.

If you're interviewing in your office, turn off all of your electronic interrupters: cell phone, pager, PDA, beeper, telephone ringer. Set your phone to send all calls directly to voice-mail. Turn off any alarms you have on your computer. Allowing an interruption to occur and then taking time to respond to it is rude to the candidate. When you allow an interruption to occur, the candidate can reasonably draw any of the following conclusions:

- You're too busy or preoccupied to spend time discussing what the candidate can bring to your organization.
- You don't actually want to hire anyone.
- You're a rude person.
- Your company is disorganized and chaotic.

None of these conclusions is helpful if you would like to hire this candidate or have the candidate refer others to you. If you want the candidate to think well of you during the interview and in the future, be respectful of the candidate's interview time and use it well.

If everyone on the interview team has a cubicle rather than a private office with a door, take over a conference room for the duration of the interview. That way, everyone has a private place to meet with the candidate. If you have an all-day interview in one room, make sure the candidate has a chance to walk around or take periodic breaks. Without breaks, people's brains can periodically turn to mush.

If you are interviewing to determine the level of a candidate's analytical or presentation skills, make sure your interviewing room is equipped with a white-board and markers. Paper is a substitute for the board-and-marker system, but is less versatile for the candidate and the interviewers alike. If you're conducting an audition, make sure the interview room has all needed equipment, such as computers, white-boards, pointers, ice water and drinking glasses, and whatever else the person being auditioned will require.

When an interview will last more than one hour, make sure that the candidate and the interviewers have access to water or some beverage. Talking for an extended period of time can make a person's throat dry and uncomfortable, which can hinder your interviewers' ability to see the candidate in the right light. Don't forget to allow the candidate a quick break after each interview, either to get something to drink, make a phone call, or to go to the bathroom.

Clarify how to handle meals.

Make it clear in advance to both the candidate and the interviewers whether you will be providing them with a meal if the interview is scheduled to stretch past normal meal times. I was once in an interview that had been scheduled to last from 9 A.M. until 2:30 P.M., and when it finally became clear to me that no one was planning to take me to lunch, I wasn't certain about what I should say or do. If I'd known I was supposed to either bring my lunch or suggest that we take a lunch break, I would have, but I was too inexperienced to say anything so I sat there as my stomach rumbled increasingly loudly.

Clarifying what the candidate can expect includes letting him or her know whether lunch is included. You can say, "We're going to be interviewing you until 1 P.M. today, but unfortunately we can't fit in a lunch break. I hope this is okay, and that you'll not be uncomfortable. Please let me know now if you think this will be a problem." If a meal will be included, ask whether the candidate has any food allergies, preferences, or dietary restrictions that you should take into consideration either in ordering food to be brought in-house or in choosing a place to go out for a meal. Decide whether you want to use lunch as a formal or informal part of the interview. For example, lunch can be used as a time to get to know the person a little, and can be an informal rather than a formal part of the interview. If you have some less-seasoned interviewers, you may want to pair a less-seasoned interviewer at lunch with an experienced interviewer. The less-experienced interviewer can watch how the experienced interviewer builds rapport with candidates and moves from question to question.

If you do choose to schedule a formal interview to encompass a meal, add enough time to the lunch-time interview slot to allow interviewers and candidates to eat the meal. I generally make such an interview ninety minutes in length, so that people have a good half-hour for the meal. It is unreasonable to expect candidates to answer questions while they are trying to eat.

Create an interview package.

Once you've decided who will interview the candidate, who will ask which questions, and for how long, it's time to create and distribute an interview package. For each candidate, I create an interview package that consists of the following components:

- the interview schedule, including interview times and room locations, and a matrix that identifies who will ask which questions and when
- a copy of the candidate's résumé
- the detailed job description

I send this interview package out to each interviewer a minimum of twenty-four hours in advance of the interview. I want the interviewers to have sufficient time to read the résumé, consider how they will ask which questions, and determine which pieces of the résumé they'll want to probe in more detail.

In advance of the interview or even during the first minutes, you may want to provide the candidate with a list of the interviewers' names and titles. Doing this gives the candidate a level of comfort and makes the process seem less intimidating. In addition, if the candidate will need to travel between buildings for interviews, make sure to give a simple map and a list of contacts to the candidate; allow enough time between interviews so that the candidate doesn't feel pressured by the schedule.

Conduct the interview.

Start the interview in a way that helps the candidate feel welcome and interested in the open position. The senior-level developer in the story near the beginning of this chapter was unhappy. So was the interviewer, because she had no time to prepare for the interview.

Make sure you don't make your candidates feel uneasy or unwelcome. Candidates have made an effort to meet with you; they want to know that you've considered who will interview them and when, and that the interviewers are ready for the interview. Make sure that the first interviewer knows that he or she is first, and that the entire team knows where they come in the sequence of interviewers and at what time.

Verify that the candidate and interviewers are ready.

Before the candidate arrives, check with all members of your interview team to make sure they are all physically present and ready to interview. If someone on your interview team is out sick, find another interviewer to replace that person or eliminate the slot from the interview and reschedule. Don't try to do this yourself five minutes before the interview; ask for help from someone in the Personnel or HR Department or ask an assistant to do the adjustments—your time will be needed elsewhere, most probably.

If the candidate doesn't appear at the appointed time, you may want to look at the résumé to see whether the candidate may be traveling a significant distance to be interviewed. At the time I schedule the interview, I usually try to ask a candidate who must travel to meet with me where he or she will be coming from on the interview date, but sometimes this detail gets overlooked and so a quick glance at the résumé can be informative. If a candidate is late, I know the interview schedule will be compromised, and so I try to find out what has gone wrong once ten minutes have passed. I approach the problem by taking the following steps:

- I check my voice-mail to see if the candidate left me a message.
- I call the candidate's contact telephone numbers, trying cell phone, office phone, home phone, or whatever phone numbers I have listed.
- I call the recruiter, if one was involved with this candidacy.

Candidates, like everyone else in the workplace, may be late to an appointment—even one so important as an interview—for various reasons, many of which have little bearing on how effective an employee he or she will be. If your candidate is late, decide on your best alternative:

- Shorten the first interview, and leave the others alone. This works particularly well if the first interviewer also phone-screened the candidate.
- Check with the other interviewers to see if they can rearrange their schedules to keep all the interviews the same duration, but at different times.
- Cancel the interview and reschedule.

When a candidate doesn't arrive and doesn't call, I eliminate the candidate from consideration.

If your interview team is ready, and the candidate has arrived, begin the interview.

Welcome the candidate.

Make sure you welcome the candidate when meeting him or her for the first time. You can welcome a person verbally, of course, but keep in mind that your facial expression also should be welcoming. A smile starts off an interview on a positive note, and makes both parties feel that the experience promises to be a good one. Some of you are probably saying, "Hey, you don't have to tell me to smile. I have good manners and I know how to behave!" Although I wouldn't disagree, I do know that sometimes it's difficult to remember to smile. Imagine the following: You're at work and you are puzzling over a work problem.   In addition, you do not feel well, and you find yourself fretting about the problem you've just discovered. You will be bound to look grim if this situation catches up with you just as you are to interview the candidate. However, the candidate won't know why you look so glum, and it's up to you to try to make a good first impression. Be friendly at your initial meeting, and you and the candidate will both feel better.

If you are scheduled to be the first interviewer, you will have a few additional responsibilities. Both to put the candidate at ease and to gather information about whether the candidate had difficulty in getting to the interview, I request that the first interviewer ask these questions:

- Did you have difficulty finding our office? (If the answer is yes, try to find out what the difficulty was. Ask, "Were the directions specific enough or was something else a problem?")
- Did you have any trouble parking? (This question is particularly relevant if your office is in a high-density area where parking is at a premium; I skip the question if parking is not an issue.)
- Would you like to use the restroom before we get started? And can I get you some coffee or tea or water?

If there is a problem, such as incorrect directions, I either note the problem to fix later, or I address it immediately. I want to make sure that even if the candidate didn't have a smooth start, his or her concerns all will be addressed.

As soon as we are settled and can begin the interview, I ask the candidate to confirm that I have his or her current résumé. If the candidate says that the résumé is outdated, something that can happen fairly frequently when an external recruiter sends the résumé, I ask for a copy of the current one. If the candidate does not have a new copy available, I ask him or her to tell me what has changed, and I mark up the résumé I have. So that the interview packets passed out to interviewers in advance are similarly up-to-date, I make a copy of the updated résumé for myself, and then ask someone else to make copies for distribution to the interview team.

You or whoever is the first interviewer may want to explain the job at the beginning of the interview day. If you've written the job description so that you can share it with the candidate, give the candidate a copy. If you've already talked about the job during the phone-screen, you may not need to discuss the job in the interview.

Once you've taken care of initial house-keeping details, you can get started. Ask questions, and listen to the answers.

Ask focused questions.

At this point, each interviewer should be ready to ask the candidate questions that relate his or her topic area to the candidate's past work. If you're a list-maker, you could make a list of questions so you don't forget which questions to ask. If you do make a list, keep the list less than a page in length, and allow yourself to move away from the list if you hear an intriguing answer.

If you don't make a list, keep the interview matrix in front of you, so you can remember your areas of responsibility for the interview.

I don't write down notes about the candidate during an in-person interview. If you feel the need to take notes, take them on paper, never on a computer. My reason for this is that when you use a computer, you have to sit behind a screen, which creates a barrier between you and the candidate. Anything that creates a physical barrier has the possibility of creating a rapport barrier as well. Note-taking during an interview increases the risk that you will destroy the conversational flow between you and the candidate. If you choose to take notes, explain to the candidate that you're taking notes because you don't want to forget anything he or she has said.

Remember, however, that if you run into legal problems with a candidate later on, notes are discoverable.[1] If you always take notes, take notes consistently. If you normally don't take notes, don't start now. If you're not sure what to do, talk to someone in your Human Resources Department or to your corporate lawyer.

Behavior-description questions require that the candidate do most of the talking in the interview. The candidate will talk about 80 percent of the time; you will talk about 20 percent. If you're not used to having the candidate do most of the talking in the interview, practice in advance with someone on your interview team to make sure you're not talking too much.

When you ask behavior-description questions, you may want to introduce them and explain that you will give the candidate time to think. You might say, "I'm using a questioning technique that will require you to

---

[1] A helpful discussion of the role notes can play in the legal process appears in Pierre Mornell, *45 Effective Ways for Hiring Smart: How to Predict Winners & Losers in the Incredibly Expensive People-Reading Game* (Berkeley, Calif.: Ten Speed Press, 1998), pp. 87, 202-8.

think about your past experiences. If you need time to think about the answer, that's okay. I'll wait for your answer."[2]

Then, sit patiently and wait. You may have to wait for a couple of minutes for the candidate to start speaking. Two minutes can feel like an eternity when you're sitting in silence, but your job is to wait. You're the one asking the questions, not prompting for an answer. Don't prompt the candidate and don't lead the candidate's answer.

If you are uncomfortable during silences, try reviewing the résumé to see what other jobs the candidate has had that are relevant to the question you're asking. Don't read your mail, the newspaper, or anything else while you wait for the answer as doing so is both discourteous and distracting: Focus on the candidate.

If you phone-screened the candidate and you're now interviewing the candidate in person, make sure you ask questions that are different from the ones you asked in the phone-screen. If you ask the same questions, the candidate may wonder whether you remember him or her from the phone-screen. If I need to repeat a question, I preface it by saying something to explain the repetition. For example, if I were interviewing a managerial candidate, I might say, "I remember from the phone-screen that we talked about your technique for handling the end-game of a project. I'd like to hear more about your last project." That way, the candidate knows that I remember the phone-screen, and that I'm looking for more detail.

You may want to explain what it was that you saw in a candidate's résumé that has prompted you to ask a question. For example: "I was impressed by your résumé. I see you worked on the Frizzit project at Whosis, developing the internal tools. How did you know what the requirements were for the tools?"

Ask lawful questions.

It is probable that every country and most regional governments have laws about what you can legally ask during an interview. Your company may have additional requirements about questions you can ask in an interview. I suggest you ask only questions that are directly *relevant to the job*, but check with your HR representative or corporate lawyer to verify what you can and cannot ask.

Questions relating to the following topics can be considered discriminatory, and may be illegal to pursue:

---

[2] For more on how to put the candidate at ease while you are explaining what you will expect of him or her, see Tom Janz, Lowell Hellervik, and David C. Gilmore, *Behavior Description Interviewing* (Englewood Cliffs, N.J.: Prentice-Hall, 1986), pp. 31ff.

- age, race, religion, sexual orientation, national origin, and the like
- height, weight, personal data
- marital status, child-care arrangements
- language fluency in English or other natural languages
- security-clearance, arrest record
- military service
- financial status, bankruptcy filings
- college or university graduation dates

You may be able to guess why asking questions in these subject areas is unwise, but don't guess. Get the facts about the laws in your jurisdiction—then devise a way to get information as needed.[3] For example, say you're interviewing a technical support person who needs to be at work during very specific hours. If you suspect the candidate has young children, you can ask, "Are you able to meet our work-hour requirements? How have you managed to meet all your work hours in the past? May I ask your references that question also?" Do not ask for details about the candidate's child-care arrangements as those details are not relevant to job performance.

Or, to take another example, if you see that a candidate has been in the military, you can ask about the projects the candidate worked on, but you cannot ask about the candidate's discharge or rank while in the military, unless the information is directly relevant to the job.

If you're not sure whether a question is legal, don't ask it. Likewise, if the question isn't relevant to the job, don't ask it.

Some topics cannot be avoided, however. If you have elimination factors, and you want to make sure the candidate can perform the work, collaborate with the interview team to develop ways to ask those questions while remaining within legal bounds.

Listen to and evaluate each candidate's answers.

During the interview, practice your active listening skills. Here are some guidelines for active listening in an interview:

---

[3] There are many good sources from which you can learn more about specific laws. See, for example, Martin Yate, *Hiring the Best: A Manager's Guide to Effective Interviewing* (Holbrook, Mass.: Adams Media, 1994), pp. 167ff.; Paul Falcone, *96 Great Interview Questions to Ask Before You Hire* (New York: AMACOM, 1997), pp. 189ff.; John Kador, *The Manager's Book of Questions: 751 Great Interview Questions for Hiring the Best Person* (New York: McGraw-Hill, 1997), pp. 192ff.; and Del J. Still, *High Impact Hiring: How to Interview and Select Outstanding Employees* (Dana Point, Calif.: Management Development Systems, 1997), pp. 114ff.

- *Stay in the moment.* Focus your listening on what the candidate is saying now. If you find yourself thinking about something else, make a note if you must, and return your attention to the interview. When I interview people, I use the physical act of closing the door to also mentally close the door in my brain to the problems I'll have to return to at the end of the interview.[4]

- *Maintain your alertness.* If you normally run from one meeting or issue to another, you may find it difficult to stay in one place for the interview. If so, consider these techniques: Dress in layers that can be added or removed so you can keep comfortable during the interview; keep water or other non-sugar beverages available to perk yourself up if you feel drowsy or inattentive; or do whatever you can to get a good night's sleep the night before the interview. There are, of course, other techniques that you can use, but these are a good start.

- *Allow candidates to complete their thoughts and sentences.* Some candidates speak more slowly than others. Some candidates continue thinking as they talk. Avoid interrupting the candidate, so that you can hear everything he or she has intended to say.[5]

- *Encourage candidates to complete their stories.* Sometimes, candidates tell a story only part of the way, thinking you're not interested or the details are not relevant to the position. If you think there's more to the story, say, "I bet there's more to that story. Tell me more."

- *Restate what you thought you heard.* Use this technique when you want to check your understanding, especially if you think you heard something that appears outlandish or even merely somewhat unusual. If, for example, a candidate states, "The project was the longest three months I ever spent," ask him or her to tell more about the experience. If you don't understand what point is being made, say something like, "I think I heard you say the role of the release engineer is to rename everyone's variables. Is that what you said, and if so, what exactly did you mean?"

---

[4] For ways to keep your attention focused, see Carol A. Hacker, *The Costs of Bad Hiring Decisions & How to Avoid Them*, 2nd ed. (Boca Raton, Fla.: CRC Press, 1999), pp. 87-88.

[5] Good listening skills can be developed, as described in Joseph Rosse and Robert Levin, *High-Impact Hiring: A Comprehensive Guide to Performance-Based Hiring* (San Francisco: Jossey-Bass, 1997), p. 179.

- *Summarize the major points of what you heard.* To close, say, "So what you liked best about your last job was pair programming. The product technology wasn't that interesting, which is why you're looking for a job. Did I get that right?"

When you actively listen, you can evaluate how the candidate answers and the quality of the answers you hear. Table 9-2 presents a checklist you can use to evaluate a candidate's answers during an interview.

Question	Interpreting the Answers
Is the candidate talking about real experience?	Analyze whether the candidate is talking about past projects and past behavior when answering the behavior-description questions, or whether the candidate is speaking in a hypothetical way: "Oh, if I were going to manage a project, I would do it like this. ..."
Does the candidate appear to have limited experience?	Sometimes, candidates have the same years of experience listed multiple times, continuing to work the same way at each job, never learning more or stretching themselves. Use your behavior-description questions to ask what was the same and what was different about each project or each job.
Is the candidate talking enough or am I talking too much?	I estimate that I spend a minute or so asking a question that takes a candidate about four to five minutes to answer. If you're talking more than 20 percent of the time in the interview, consider why. Are you asking close-ended questions? Are you leading the candidate to the answer you want to hear?
Is the candidate hijacking the interview?	Is the candidate taking over the interview and not answering the questions you want answered? Sometimes, the candidate is a talker, and you may need to interrupt to restate your question or to bring the answer back on track. If you have to interrupt the candidate more than once, note the kinds of answers you interrupted the candidate on. You'll want to compare notes about the candidate with the other interviewers to see if they also had this kind of problem.

Table 9-2: Questions to Ask Yourself About a Candidate's Responses.

While listening, you may decide that you need to guide the interview in another direction. Replanning the interview is fine. I recommend that you check the interview pace about halfway into your time and make sure you're covering the topics you need to cover. If you're behind schedule, see whether you can focus your questions more tightly. If you're ahead of schedule, try to encourage the candidate to speak at greater length about specific experiences.

Answer the candidate's questions.

Leave the last two-to-five minutes of the interview open for the candidate to ask you questions. Some candidates will have questions; some won't. If the candidate has questions, answer them to the best of your ability, always telling the truth. If the candidate has no questions, suggest that he or she take your card to call you later if any questions crop up.

Deliver the candidate to the next interviewer.

Before taking the candidate to the next interviewer, offer the option of a visit to the restroom or of getting something to drink. Because I don't like to drop a candidate off in someone's office if the interviewer isn't there yet, I stay with the candidate until the interviewer appears and is ready to start the interview. If I must wait with the candidate for five minutes into the next interview slot, I call the missing interviewer's telephone or pager rather than wait any longer. I once had to track down an interviewer who was focused on solving a problem in the lab and had completely lost track of the time. If I hadn't stayed, we would not have known that the interviewer had forgotten about the interview in time to rectify the problem.

After the first interview has been conducted, subsequent interviewers should not try to make small talk with the candidate. Start with the interview questions, and make the most of both your and the candidate's interview time.

Conduct group interviews sparingly.

I'm not a fan of group interviews. Most candidates find it intimidating to sit across the table from three, four, or five strangers. Inevitably, one person asks most of the questions and other interviewers have little opportunity to follow up on answers they found interesting. I experienced first-hand how ineffective a group can be when I was interviewed by a team of interviewers, two of whom jockeyed for the lead position and ended up asking each other questions. In roughly an hour, I talked for only about ten minutes. I decided that if the team members were that ineffective at interviewing, they couldn't be very effective working together. I withdrew my candidacy based on that one experience.

In general, group interviews waste most everyone's time. You may think you're saving time because you've reduced the total interview time, but you've also limited the information each interviewer can receive about the candidate.

I do, however, sometimes use two-person interviews. If I'm training a junior-level person on interview techniques, and we haven't had enough time to practice, I'll ask a senior-level person to co-interview a few candidates and then coach the student interviewer for the next few interviews. I also ask the primary interviewer to explain the student's presence to the candidate.

If you decide to try panels or group interviews, limit the number of people on one panel to three or four. Decide before the interview who will ask which questions when. Schedule the panel interview late on the first interview day, or preferably, on a future date, after the candidate has already met some members of the interview team. Allow more time for a panel interview than you would for a one-on-one interview—two hours is about right. Make sure the panel members understand who will take which role, who will ask follow-up questions, who will lead the discussion, who will ask for examples of previous work, and so on. Panel interviews require considerably more preparation than solo interviews in order to make them worthwhile.[6]

End the day of interviews.

Just as the first interviewer has specific responsibilities, so does the last interviewer. Whoever closes the process should do three last things: Thank the candidate for coming in, explain that the hiring manager will get back to him or her with status within a day, and escort the candidate to the same entrance he or she arrived at.

If you work in a company that requires the person who signed a candidate in also to sign him or her out, make sure the initial interviewer is available at the end of the last interview, and don't ever leave a candidate at the door of one of many buildings in a multi-building complex, where he or she would have to find the way across a large campus back to the original entry point. These three acts may seem simple but they are easy to overlook at the end of a long day.

By showing respect to both the candidate and to the other interviewers, you make the interviewing environment comfortable. Likewise, by providing members of the interview team both with the materials they need and the time to review those materials, you show them you respect them. It is also important to show that you respect yourself. One sure way to do this is to prepare the candidate and the interview-team members for the experience, and to make decisions quickly. Interviewing can be stressful for candidates and interviewers, but the more consideration you show for all the people involved, the more successful your interviews will be.

---

[6] Group or panel interviews are discussed in Lou Adler, *Hire with Your Head: Using Power Hiring to Build Great Companies*, 2nd ed. (Hoboken, N.J.: John Wiley & Sons, 2002), pp. 162ff., 286-87.

Case Study: Walker Software

Dirk is continuing to recruit for a developer. He has drawn up an interviewer matrix, as shown below.

Interviewers	Dirk	Vijay	Susan	Steve	Sam	Everyone
Time	8:00-8:45	8:50-9:35	9:40-10:25	10:30-11:15	11:20-12:05	12:10-12:55
Location	Dirk's office	Conf A	Conf A	Conf A	Conf A	Dirk's office
Question Areas						Meet to evaluate the candidate
Design Skills				x	audition	
Collaboration and Team-work Skills	x	x	x			
General Problem-solving Skills	x				x	
Decision-making Skills		x	x			
Testing Skills		x		x		
Technical Process and Methodologies	x			x		

POINTS TO REMEMBER

- Select who will be on your interview team. Make the team cross-functional if the candidate will need to work across corresponding functional areas.
- Ask each interviewer to define his or her interview's duration, but make the interviews long enough for people to gather data about the candidate.
- Choose an interruption-free, comfortable place to interview.
- Create and send to each interviewer the interview package at least twenty-four hours before the interview. Follow up on the day of the interview to make sure everyone is still available to interview.

- Develop your own behavioral questions, and encourage everyone on the interview team to prepare unique questions. Actively listen for answers.
- Ask questions that comply with the law.
- Be considerate of the candidate's time and effort.
- Be considerate of your interview team's preparation and interview time.

10

Following Up After the Interview

Marty: "So, what did you think of the first candidate?"

Jane: "Oh, he was great. Offer him a job."

Sue: "Hey, wait just a minute. The first guy today wasn't a guy, he was a woman. Well, she was a woman. You've got the candidates confused."

Marty: "Um, no, I meant the first candidate yesterday—Jeff. I wasn't so sure I liked him, but we need to talk about everyone, so let's start with Jeff."

Jane: "Jeff was first? I talked to Simon first yesterday."

The opening scenario illustrates how easy it is, even for experienced interviewers, to get caught up in a limbo of confused discussion when conducting a two-day interviewing blitz. Jane, Sue, and Marty think they are talking about the same candidate, but each interviewed a different candidate "first." This problem would have been avoided if Marty had identified *by name* the candidate he wanted to discuss first, but confusion can also arise when interviewers have seen so many candidates in a short period of time that they begin to blur one candidate's personal strengths or weaknesses, professional qualifications, and suitability for the open job with another's.

Instead of allowing confusion to take root during two days of interviews (or whatever time frame I am working within), I schedule the interview so that the interview team sees only one candidate during each half-

189

day period. Then, I bring all members of the interview team together for fifteen-to-thirty minutes after the candidate leaves and before the next candidate arrives. The purpose of the meeting is to compare notes on the candidate, and then decide whether to keep the candidate on the list of potential hires, to reject the candidate, or to make an offer to the candidate. If you have only one open position, then interview a few candidates before you make the final decision—don't offer the job to the first candidate the interview team likes.

If the team members can't decide which action to choose within fifteen-to-thirty minutes, they may not have been given enough information either about the position or about the candidate. Listen to whatever reasons they give for why they can't decide, and then address that difficulty. You may need to expand upon the details you have recorded in the job analysis, or you may need to revise the interview plan. A third possibility is that the candidate has not presented himself or herself to anyone's complete satisfaction or has been vague about work experience or other requirements essential for the job. In such a case, my advice is to reject the candidate.

Meet immediately after the candidate's last interview.

Make sure you elicit feedback from all interviewers as soon after the interview session completes as possible. One of my colleagues told me about his disastrous experience when he failed to follow this strategy: "I once hired someone before all members of the interview team had completed interviewing the candidate. I jumped the gun because every one of the early interviewers was really enthusiastic and the final interview—the technical interview—wasn't scheduled until later in the week. When the last interviewer held the technical interview, she told me not to hire this person because the technical depth wasn't there. That hiring was a big mistake on my part. The person didn't work out and we had to take action later to terminate employment."

That colleague's unfortunate experience is worth keeping in mind: An important part of consensus-based interviewing is to pay attention to *each* interviewer's reaction to the candidate. So, even if you're in a rush to hire people, make sure you hear feedback from the entire interview team. And, when you listen to the feedback, review what people were supposed to discuss with the candidate, to make sure interviewers covered the areas they either elected to cover or were assigned.

The sooner you bring interview-team members together after they have met with a candidate, the more details they will remember, and the more valuable the discussion of a candidate's merits will be. An additional advantage of holding immediate post-interview meetings is that they enable you to use limited consensus based on an exchange of accurate information about the candidate—resulting in a situation in which everyone in the group has a voice in the decision and can abide by the decision, but in which the person to whom the candidate will report can have the greatest say.[1] It does *not* mean that everyone has to be happy with the decision. As the hiring manager, the hiring decision is yours, but your interview team can supply you with valuable information so you don't have to make the decision in a vacuum.[2] One other technique I've found to be advantageous is to invite internal or external recruiters to the meeting, even if they only can participate by means of a conference call. After hearing details about what my team members and I like or dislike in a candidate, recruiters often can choose better-suited candidates to send to me in the future.

The immediate post-interview session is not always feasible, however, particularly if you're interviewing until late in the day. In such a situation, I ask interview-team members to make notes immediately after the interview and then ask them to meet with everyone at the beginning of the following day. It is better to hold the assessment meeting at the beginning of the day with refreshed team members rather than with hassled and harried staffers who are trying to get out of the office at the end of a long day.

Hold the meeting in a private space.

If you have an office big enough to hold everyone, you can conduct the post-interview meeting there. If not—or if you don't have sufficiently private space—plan to book a conference room or another suitable empty space.

Facilitate the meeting.

When you facilitate a limited-consensus meeting, you're looking for information. Ask, "What makes this candidate a great fit? What makes the candidate not fit? What further actions should we take?"

---

[1] Factors to consider during the final-decision-making process are given comprehensive treatment in Carol A. Hacker, *The Costs of Bad Hiring Decisions & How to Avoid Them*, 2nd ed. (Boca Raton, Fla.: CRC Press, 1999), Ch. 15.

[2] In her chapter on developing and keeping talent, Sherrie Gong Taguchi presents a solid discussion of feedback-gathering techniques. See Sherrie Gong Taguchi, *Hiring the Best and the Brightest: A Roadmap to MBA Recruiting* (New York: AMACOM, 2002), pp. 247ff.

Consider asking your interview-staff members to take five minutes as soon as they are done with their interview to write down what makes the candidate suitable or unsuitable for the company, to note whatever concerns they have about the candidate, and to record data about the candidate. When you ask your interview-team members to jot down their thoughts, you help to assure that they don't come into the evaluation meeting cold without taking the time to think about the candidate.

Once I have everyone together, I ask people to show their reaction with just their thumbs: A thumb up means, "I like this person. Hire him (or her)!" A sideways thumb means, "I can live with this person." A thumb down means, "I object to hiring this person." I make note of how many people voted for or against hiring, to be compared to a second vote that I solicit later in the meeting.

If I have an interview team on which even just one person has an overpowering personality, I ask all the interviewers to shut their eyes when they vote with their thumb. If this approach seems awkward or even corny to you, an alternative is to issue to each interviewer three slips of paper to use as ballots, on each of which is written a different word selected from either YES, NO, or MAYBE. This will allow people to vote their perspective anonymously. However, when I'm not worried that outspoken people might take over the meeting, I use the thumb-vote technique to elicit each interviewer's honest thoughts about the candidate.

I ask for thumbs rather than opinions because I've found that some junior-level or less-opinionated members of hiring teams feel pressured to agree with other interviewers. Since I want everyone's honest opinion, I ask for thumbs first. Once we know the state of the thumbs, I can ask more questions. Most people who participate in this kind of vote don't feel pressured by their peers into changing their vote.

Learn the reasons behind each thumb-down vote.

Before going any further with the assessment meeting, I ask thumb-down voters to explain their reasoning. I never assume that I know why someone has given a thumb-down vote to a candidate, so I ask for clarification. One thing I want to be confident of is that no interviewer has personal issues about the candidate or is influenced by an unfounded bias.

I listen to people's comments and then ask whether other interviewers are in agreement. For example, if one interviewer gives the reason, "The candidate never answered my questions directly," I ask three follow-up questions, the first two of which I ask of the thumb-down interviewer, and the third I ask of the whole group:

- *What questions did you ask?* As the hiring manager, I need to verify that the interviewer asked appropriate questions, and I want to assess the circumstances under which the interviewer thought the candidate didn't answer. Sometimes, an interviewer's questions are too vague—or possibly too specific—for some candidates to answer in the manner the interviewer anticipated. If so, I have an interviewer problem, but I do not necessarily have a candidate problem.

- *What did you see or hear that led you to that conclusion?* In this case, I ask, "Which questions did the candidate avoid?" If someone else asked a similar question and was happy with the answer, then I want to know how early in the interview day the dissatisfied interviewer asked the question. Sometimes, a candidate who has not been exposed to behavior-description interviewing questions will need a long time to think before he or she can answer the question well. If you interview candidates like that, you'll find that answers given in their later interviews are crisper than their earlier ones. If this is the case, I have a different decision to make: Can I hire someone who needs a little more time to think?

- *Did any other interviewer have a similar experience?* If several people had the same experience with a candidate, such as feeling that the candidate avoided answering a question, I can assume the candidate couldn't or didn't want to answer the question.

Interviewers may dislike candidates for other reasons, saying, "I didn't see any evidence of problem-solving in the project," or "I can't believe this guy was a technical lead." Help your interviewers to elaborate on any negative reaction by asking them the following questions: Which behavioral questions did you ask? What answers did you hear that led you to that conclusion? Did anyone else hear the same thing?

As an interviewer answers my questions, I encourage him or her also to tell me the wording the candidate used to answer the behavioral questions. After all interviewers have repeated the answers given them by the candidate, we have information that we can compare between interviewers. One interviewer may have done a more complete job of drawing out information than another, and this process of comparing data makes evaluation of a candidate less subjective.

Sometimes, the reason someone gives for not liking a candidate will be difficult to articulate. A thumb-down vote may be explained, "I just have a gut feeling about this person not being right." If your interviewers tell you they have concerns about a candidate but cannot tell you why, probe several specific areas:

- Was there evidence of the candidate's reliability?
- Did the candidate talk about taking responsibility?
- Did the candidate give examples of taking initiative?
- Did the candidate behave in any way to make the interviewer feel uncomfortable?
- Did the candidate say something that would make the interviewer reluctant to work with him or her—and if so, what was it?

Elicit a qualitative answer from your interviewer before you determine whether to go back to the candidate to ask a second round of questions or to question the candidate's reference providers about the problem. Ask interviewers specific questions such as the ones listed above to help them articulate what triggered their gut feeling about the candidate.

Sometimes, even after you've probed for specific reasons, an interviewer still won't be able to explain why he or she is uneasy about a candidate. If everyone else is enthusiastic about the candidate, you can choose to bring in the candidate for another interview or audition, or you can pass on the candidate. Whatever you decide, don't ignore gut feelings; just don't base your decision entirely on them.

Understand the thumb-sideways responses.

After examining the reasons given by the thumb-down interviewers, I ask the thumb-sideways interviewers to share their reasons for being noncommittal: "What would it take for you to be enthusiastic about hiring or working with this person?" Answers to this question can help *all* the interviewers to assess what was positive as well as what may have been lacking in the candidate. If, from this discussion, you discover information about what the interviewers regard as essential technical and non-technical qualities, preferences, and skills needed for the job, you can use the information to compile a list of questions about essential attributes and qualifications to use as you phone-screen and interview candidates in the future.

Understand the thumb-up votes.

Finally, I invite the thumb-up people to give me their reasons for liking the candidate. I ask, "What was most exciting about the candidate to you?" That question helps people articulate what they found positive about the candidate. It is probable that different interviewers will identify different characteristics; hearing about what each interviewer found positive—and why—will help you make a better hiring decision.

Revisit the thumbs one more time.

At the end of the thumb-vote discussion, I ask if anyone's changed his or her mind. Not infrequently, some interviewers change their decision about the candidate after hearing what other interviewers found positive or negative. I compare the new tally with the count I noted earlier.

Use limited consensus to make a decision.

If everyone votes thumb-down, you usually can eliminate the candidate from further consideration.

If most people are just lukewarm (thumbs sideways) about this candidate, review your options by asking yourself the following questions:

- *Do we have additional candidates still to be interviewed?* If so, and you think you will find a better candidate, don't proceed further with this candidate. If you have few potential candidates, you may want to bring the person back for a second interview.
- *Is this a first-round interview?* If so, then it's probably worth bringing this candidate back for a second interview.
- *Is this a second-round interview?* If you already have had two interviews with this candidate, and many people are still neutral, it is usually best to decide not to pursue the candidate and to keep looking, unless you're desperate for a person to fill the position and you believe a lukewarm hire is better than no hire.

If everyone votes either thumb-up or thumb-sideways, you have achieved limited consensus that this candidate is worth pursuing, either by bringing him or her back for another interview, or by checking references and making an offer (assuming, of course, that the references provide positive feedback).

If even one person votes thumb-down in the second call for votes, I generally don't pursue the candidate. I take this somewhat severe approach because I know that my staff members will need to work with whichever person is hired. If there's a valid reason that one of my current staff members doesn't like the candidate, then I believe I need to drop that candidate.

You can also use limited consensus as a way to improve team morale and to reinforce your perspective on hiring. The following story is about a manager who did just that:

> "Hiring by consensus has helped me twice in significant ways. The first time was when a member of my team recognized a candidate as someone she had worked with before. She remembered that he had been allowed to quietly leave a previous job on which he'd had some serious performance problems. She did not want him to bring those problems to our team. When she realized that she could speak freely, I received two benefits: I avoided a bad hire and reinforced a great employee by showing my confidence in her judgment.
>
> "Another time was with a candidate who didn't have the 'usual' credentials for our team, but had a tremendous amount of experience. One of our team members was very impressed by the candidate and, after discussing pros and cons with the group, convinced the rest of us to hire the candidate. He turned out to be one of the best testers I have ever hired."

Limited consensus is a tool for you, the hiring manager, to use to collect data to make a decision about the candidate. Use the meeting to gather your data. If you choose not to be bound by the group's decision, tell your interviewers in advance that you're gathering information from them and that you will make the final decision. If you still want to pursue the candidate, do so.

Use follow-up forms with care.

After an interview, some hiring managers may be required to fill out follow-up forms issued by their company's Personnel or Human Resources Departments. Such forms constitute important legal documents for the company, but they are not a substitute for follow-up meetings.

A True Story

Darren sent e-mail to all members of his interview team asking them to gather in his office at 11:45 A.M. to discuss the candidate as soon as Tony, the last interviewer, was done with his interview. Darren had considered having the group meet in Conference Room A, but he was concerned that the candidate, Stan, would be nervous if he saw all the interviewers descending on the conference room as he was leaving.

When Tony popped in to Darren's office a couple of minutes before noon, everyone else was already present. "Okay, we've got everyone," Darren said. "Did you write down what you liked and what you didn't like about Stan as a candidate for the job?" Everyone except Tony nodded yes.

Darren turned to Tony. "Do you need a couple of minutes to write things down?" "No," Tony quickly responded. "I just finished interviewing Stan, so I'm okay."

"Okay. Could everyone now please give me a thumb-vote signal to show me what you think," Darren asked, as he stuck his own thumb up.

Without hesitation, Angus, Don, and Tony each held a thumb up. Only Susan held her thumb down. "Hah!" Tony said. "You're outvoted, Susan, and—"

"Now just a minute," Darren interrupted. "Susan, why is your thumb down?"

"Well, I just didn't like Stan in terms of this job," Susan said. "When I asked him about the makeup of his team, he told me that that he only had men on his team. He was comfortable working with men, but he wasn't sure how to work with women."

"Did you ask him more questions about how he's worked on teams in the past?" Darren asked.

"Yes," Susan said. "He told me that he'd worked on small teams and small projects, and that all his teammates were pals of his. That was his word, pals. I'm not a pal, I'm a colleague."

"I didn't hear any of that in my interview with him," Tony stated. "But I wasn't looking for it either."

"I thought I heard a little of that," Angus said. "It didn't seem like a big deal at the time, though, and I didn't think to probe deeper."

"Well, I think we'd all agree that that kind of bias is a problem," said Darren. "We have another candidate coming in tomorrow, so let's see what happens. I'll tell Stan that we're still interviewing, and that I'll get back to him tomorrow. If we decide to interview him again, we'll ask more questions about how he works in teams."

The next day, all members of the interview team met in Darren's office again, this time after interviewing George. "Okay, what did you each think about George?" Darren asked.

This time, Don was the only one with his thumb down. He quickly explained, "When I asked George about his test development for his code, he said that he hadn't done much. He's not used to working the way we work, and I'm concerned about that. On the other hand, when I auditioned him for his design skills, he was great. I think he can learn to test the way we do, but it's not how he works now."

"I'm willing to live with that if the rest of you are," said Darren. "What about his other skills, including teamwork?"

"I liked his skills better than Stan's," Tony commented. Within minutes, everyone agreed that George had most of the technical and teamwork skills he would need to work in the group.

"So, can you all live with teaching George how to test his code?" Darren asked. "He would have a learning curve, and I don't know how long it will be. Knowing we have to teach George how to test his code, let's have a show of thumbs to tell me how you feel about him."

Everyone held a thumb up.

"Well, that's it then," Darren said. "I'll contact George's references, and if they check out, we'll make an offer. Thanks, everyone."

If you and your interviewers are required to fill out follow-up forms, ask your personnel or HR representative what happens to the forms after you surrender them. If you know how the data will be used, you and the members of your interview team can do a better job of filling such forms out. Sometimes, the forms are filed as the "official" Equal Employment Opportunity record of an interview. If so, make sure all comments are both factual and free of flip or sarcastic comments.

If your interviewers fill out these forms thoroughly and promptly, you can review them with an eye to understanding whether you phone-screened effectively and whether you have described the job correctly. If interviewers time after time fill out the form indicating, for example, "Candidate does not have enough technical skills to solve our kinds of problems," you can ask the interviewers what technical skills they are looking for. Then, you can add questions pertaining to those skills to the phone-screen.

If you use follow-up forms, make sure that questions such as the following are addressed on the form:

- Should we hire this person?
- What makes you think this person is a good candidate for the open job and for our company?
- In what ways does this person fit or not fit the job?
- What concerns, if any, do you have about this candidate?

Although a follow-up form provides you with less detail than you would learn from holding a meeting about each candidate, it can be informative nevertheless. Be wary, however, of forms that only elicit yes/no answers to questions, as they will not supply you with enough information to enable you to make an informed decision or to assess the effectiveness of your initial phone-screen and interview questions.

Even the most conscientiously filled-in form probably won't provide adequate feedback about a hiring effort, but if your organization requires you to fill out a form, fill it out. If nothing else, your personnel or HR representatives will appreciate your compliance and may be more responsive to you when you approach them for assistance in the future.

In the United States, companies are expected to maintain—for at least one year—a record of all résumés received and candidates interviewed.[3] Because these records may be used as legal documents, they should be treated as such. If your company will use your follow-up forms as legal documentation to back the decision not to hire a candidate, have them

[3] A clear explanation of the Age Discrimination in Employment Act's requirement that employers preserve such records appears in Robert W. Wendover, *Smart Hiring: The Complete Guide to Finding and Hiring the Best Employees,* 2nd ed. (Naperville, Ill.: Sourcebooks, 1998), p. 30 and Appendix B.

reviewed and approved by your corporate lawyer before putting them in the record.

Tell the candidate what to expect next.

Whether you will invite the candidate back for another interview, or you will proceed to reference-checking, or you will tell the candidate that he or she does not have the requisite skills for the open job, you need to be explicit about what the next steps will be. As soon as you have advised the candidate about the next steps, be sure to tell recruiters and personnel or HR representatives who are working with the candidate where he or she stands, as described further below.

Any external and internal recruiters, personnel staff members, and HR representatives who have worked with you and the candidate all need to know what you think of the candidate. Use this opportunity to explain what members of the interview team liked about the candidate and what they didn't like. When you take the time to explain the pros and cons, your professional contacts will know more about the kinds of people you need and will be able to provide better-suited candidates in the future. Make sure, however, that both you and your contacts know who will let the candidate know the decision.

Even if you're not sure what the next step should be, you still need to keep the candidate informed. Diplomatically tell the candidate the truth, no matter what that is. If you're not sure you want to schedule another interview or move forward to contact the candidate's references, but you don't want to reject the candidate outright, you can explain why you would like more time before reaching a decision, saying something like the following:

- "We're interested in you, but we have already arranged interviews with several other people, so we're going to finish our interview process before we make a decision. We'll get back to you early next week."
- "We liked what we saw in you yesterday, but we need to re-assess our job requirements to verify that your skills and experience do satisfy a majority of the requirements."

Comments such as the preceding buy you more time. However, if you're not interested in pursuing the candidate further, be truthful—but not unkind—perhaps saying, "Thank you for interviewing with us. We're not ready to go forward with you now, but we will keep your résumé and application on file. In the meantime, we wish you every success in your job search."

When you want to proceed with a candidate, state your interest, saying something along the lines of the following: "We're very interested in you, and would like you to come back for another interview if you are interested in doing so." Or, if your next step will be checking the candidate's references or making an offer, tell this to the candidate.

POINTS TO REMEMBER

- Meet with all members of your interview team as soon as possible after the last interview.
- Make sure all interviewers have a chance to express their perceptions of the candidate without feeling intimidated by strong personalities in the group.
- Let candidates know where they stand as quickly as possible.
- Use limited consensus as a way to build the team; don't hire by management fiat.

Part 4

Bringing In the Candidate

You've found a candidate that you and your hiring team like, and you're ready to check references and make the offer.

Be prepared to speed through this part of the hiring process. The less time you take to check references and make an offer, the more likely the candidate is to agree to your offer. The faster the candidate agrees, the sooner he or she can start. But you want to do as thorough a job on reference-checking as possible, even in a short period of time. So how do you do this?

First, set as your goal the job of checking references within twenty-four hours of the candidate's last interview. Once you've checked references and decided the candidate is right for the job, do all you can to make it easy for him or her to say yes. One way to make things easier for the candidate as well as yourself is to prepare the offer while you check references. Take care to devote sufficient time to completing all paperwork associated with the offer (including getting any necessary signatures or sign-offs), so that the offer is ready when the reference-check is complete.

To help you better understand what tasks are associated with bringing in the candidate, Chapter 11, "Checking References," details techniques for fully utilizing references. Chapter 12, "Creating, Timing, and Extending an Offer," provides pointers to use as you near the end of the hiring effort.

Checking References

Middleware Manager: "Joe, I need to talk to you about your bug-fix documentation. It's not detailed enough for the testers to figure out how to test your fix."

Programmer: "Oh, that's what my last manager said, too. I know I should provide better documentation but I hate dealing with details."

Middleware Manager: "Well, you may hate details, but details are absolutely necessary when you document your fixes.

Had the manager in the opening story checked this employee's references thoroughly before hiring him, she undoubtedly would have learned that he was not detail-oriented. Even more important, she probably also would have learned that the documentation he had delivered on the previous job was not satisfactory to that manager. Knowing of these performance shortcomings, she probably would not have hired him.

There is a lesson to be gleaned from the story: Always check a candidate's references to confirm that he or she is competent at the tasks you will need performed. Reference-checks can help prevent you and your colleagues from being unpleasantly surprised when the candidate has been hired and starts work as an employee. Reference-checks help you complete your picture of the candidate.

If you feel pressured to make an offer to a candidate who you believe looks promising—pressured possibly because you suspect that the candidate has other job offers—make an early offer *contingent on good reference-checks*. Here is an exchange Louisa, a test manager, had during a telephone conversation with a job applicant's previous manager.

"Hello, Tom. My name is Louisa Seniority and I am the test manager at Flashy Software. James J. Applicant has given me your name as a reference, and I wonder if you have a couple of minutes to tell me a bit about him?"

"Absolutely."

"So, would you hire James again?" Louisa asked.

"Positively," replied Tom without a moment's hesitation.

Louisa needed more detail. "What made James such a valuable employee?"

"Oh, he was wonderful at test planning to the most detailed degree," Tom explained. "I could always depend on him to create test cases that other people could execute."

"Oh, great," Louisa said. "How about exploratory testing? Was James good at that as well?"

Tom thought for a moment and then responded, "Oh, we don't really do that here. I have no idea how he would be for that kind of job. Why would he need to know that?"

Louisa explained, "It's one of the jobs he'd need to perform here. We need someone who is good at exploratory testing. Well, I guess James is not quite right for the job, but I do thank you for your time, Tom."

Louisa's "guess" wasn't justified. Because Tom doesn't know about exploratory testing, Louisa should not use him as a reference for James's ability to perform that part of the job. The conversation was useful nevertheless as it confirmed that Tom considered James to be highly qualified as a test planner. Because Louisa is looking for someone who can do both test planning and exploratory testing, she doesn't have enough information to conclude that James is or is not a qualified candidate. However, she does have a great reference for James's ability to plan tests, so she's partway through her reference-check.

Check all offered references.

Few people who are responsible for hiring need to be reminded that it's worthwhile to check references, but sometimes reference-checkers fail to use the process to full advantage. I recommend checking *all* references and using a portion of the time for the following:

- *Verify that the information you heard in the interview is accurate.* Sometimes, interviewers hear what they want to hear, rather than what the candidate says. If I have two interviewers who thought they heard different things, I check the accuracy of the information with each reference. I've used reference-checks to confirm, say, an applicant's position on a project, the level of technical contribution, and how well the person worked as a member of a team.

- *Ask what it was like to work with the candidate.* Sometimes, even after my interview team and I have finished interviewing a candidate, I have some lingering questions about what it's like to work with him or her. By asking the reference to describe his or her experience, I can get a better sense of the candidate's work habits, strengths, and weaknesses.

- *Confirm a candidate's fit in the organization.* As a hiring manager, I have to deal with the consequences of hiring people who don't perform well or who don't fit into the group. To minimize the likelihood of hiring the wrong candidate, I use the reference as a checks-and-balances mechanism. I ask for as much detail as the reference is willing to give me about every aspect of the candidate's previous responsibilities, professional preferences, and on-the-job achievements. A word of advice: If you've hired more than your share of people who haven't worked out, look back at your job analysis, interview questions, and reference-check questions, and see whether you could have asked the candidate different questions, or asked the reference questions differently.

When you are the hiring manager, it's in your best interest to perform the reference-check yourself.[1] This is because, whether the candidate works directly for you or is matrixed to other projects, you are responsible for how well the candidate will fit into the organization. Therefore, it is up to you to know what your own and your team members' preferences are—and how well the candidate fits those preferences.

Although candidates select the people they list as their references—and are not likely to name a person who might give an unfavorable response—you usually can obtain a reliable sense of a candidate's character, competence, and worth from references. It is true that some references are unable to give you anything other than dates of employment and final salary because of their company's policies or perhaps because they have nothing very good to say, but most references will answer honestly as to whether

---

[1] The argument that the hiring manager should do the reference-checking is nicely made in Carol A. Hacker, *The Costs of Bad Hiring Decisions & How to Avoid Them*, 2nd ed. (Boca Raton, Fla.: CRC Press, 1999), p. 123.

they would rehire the candidate. If you run into references who don't answer your questions as fully as you'd like, you might ask the candidate for alternative references.

Develop your list of reference-check questions.

Reference phone calls are similar to interviews. Use a combination of open-ended and close-ended questions so you can see a clear picture of what the reference thinks of the candidate's past performance, ability to work with others, initiative, and other such matters relevant to the open position.

I use the following as an outline for telephone calls made in the context of a reference-check:

1. facts (including verification of information the candidate reported verbally in the interview or listed on the résumé)
2. issues (including information about any room for improvement in performing the job, difficulty of assignments, positive attitude, timeliness)
3. mystery areas (including questions addressing topics not answered to my team's or my own satisfaction in the interview or pertaining to ambiguous areas)
4. rehire or re-fire
5. closing comments

I always ask the same basic questions for reference-checks, regardless of the position I want to offer. I do this for two reasons. First, I want to get a true picture of how each candidate relates to each other candidate—an "apples to apples" comparison. Second, I want to ensure that all people named as references are providing a consistent picture of a candidate's strengths and weaknesses. To the set of basic questions, I add questions specific to each candidate, as needed, to resolve any ambiguity left over from the interview.

When you develop your reference-check questions, make sure the questions relate to the job. I write a reference-check script, just as I write a phone-screen script, on which I keep a record of the answers I hear about a candidate.

Questions I have found useful when asking for a reference for a technical candidate are shown on the form below. I've added explanatory comments as parenthetic matter to share tips I've learned over the years:

Candidate name: _____   Date:_____

Reference name: _____   Date: _____

Reference job title: _____

Reference telephone numbers: _____

To report to:

1. **Where and in what capacity did you work with the candi-date?** (This close-ended question establishes the reference's relationship with the candidate. If the reference cannot comment on the candidate, then ask to speak to someone in the Personnel or HR Department to confirm dates of employment and salary. Then move on to the next reference.)

2. **How long did the candidate work for you?** (If you're calling the candidate's most recent manager, use this close-ended question to establish how long the reference was the candidate's manager.)

3. **How long have you known the candidate?** (Use this close-ended question to determine whether the reference is a colleague of long-standing or someone who does not know the candidate well. If all the references describe relatively new relationships (six months or less), I ask the candidate for the names of people who have had a longer work or academic relationship. Even a recent college graduate with little or no work experience is likely to have had a relationship of six months or longer with a classmate, academic advisor, or professor who would be willing to serve as a reference. Although a candidate's lack of a long-term reference may be caused by the fact that some employers do not permit employees to give references for colleagues or even subordinates, one reason for a candidate's lack of a long-term reference could be that he or she has been so difficult to work with that no one is willing to say anything at all.)

4. **How would you describe your working relationship?** (This open-ended question encourages a discussion of how the candidate performed when under supervision.)

5. **Describe the most recent project the candidate worked on with you.** (This open-ended question encourages quantitative and qualitative information that describes the complexity of the candidate's work, and how well the candidate did.)

6. **What issues did you have with the candidate's work?**
    (Make sure this is an open-ended question so that you can
    follow up on any concerns you have from the interview. If
    any of your interviewers had negative feelings about the
    candidate, use this question to determine why. Sometimes, if
    I have doubts from the interview about the candidate's
    ability to complete work, or the candidate's estimating
    expertise, for example, I follow up and say, "Did anyone ever
    complain about this candidate's ability either to complete
    work or to estimate work?" This question has special impor-
    tance to me: Many years ago, I fired a junior-level developer
    because he was unreliable about coming to work and conse-
    quently had difficulty finishing his work within the time
    allotted. A few months after I fired him, he gave my name
    (unbeknownst to me) as a reference for another position.
    The hiring manager called me and asked, "Did you have any
    issues with the candidate's work?" I felt uncomfortable
    answering with what I knew could damage the developer's
    chance for a fresh start, and so I gave a noncommittal
    response, saying, "I have no comment." The hiring manager
    persevered, "I have information from my interviewers that
    this fellow may have trouble managing his time and activi-
    ties at work. In fact, he told the interviewer that the fol-
    lowing happened at your company." The hiring manager
    described the situation, and then asked, "Can you verify
    that the situation occurred?" Since the hiring manager
    already had information indicating the problem, I provided
    confirmation. Be forewarned, however: If you hear a no-
    comment statement from a reference, it may be a guarded
    warning meant to alert you to a previous problem. Getting
    information from someone who has said "No comment"
    takes skill and tact. I sometimes ask, "If the candidate has
    any areas needing improvement, what would you say they
    are?" Just as in the interview, I ask behavior-description
    questions, focusing the reference on how the candidate per-
    formed required work, perhaps requesting, "Tell me about
    the feedback you gave the candidate, and, if you had specific
    issues, what actions did you and the candidate take to
    resolve them?")

7. **If I were the candidate's manager, what advice would you have for me?** (This open-ended question asks people for advice and gives a reference a great way to provide a candid response. As an alternative question, you can also ask "What would this candidate need to do to be more successful?")

Other questions you may choose to ask, particularly when hiring a staff member, can include the following five:

8. **How quickly did the candidate learn about the product or the product line?** (If product-line understanding is important to you, ask this question, which you can follow up with other questions about how well the candidate understands the industry. Although you may not need this question, much of my hiring has been in organizations where it was critical for new staff members to quickly learn about the product. If something else is critical for a new hire to learn quickly, ask that question here.)

9. **How quickly did the candidate integrate with the rest of the team?** (Start with the close-ended question to learn whether the candidate did, in fact, integrate with the rest of the team. Then you can follow up: "Were there people with whom the candidate found it especially easy to work? Were there people with whom it was harder for the candidate to work? Tell me about the candidate's communication style." Especially if you work in smaller teams, you'll want to know whether the candidate has sufficient social skills to work with other people. You can choose to hire people without social skills, but if you at least know from the start that they lack them, you can compensate for the shortcoming some other way.)

10. **Why will or did the candidate leave?** (Use an open-ended question to learn about the candidate's motivation for leaving the previous job. Sometimes, the reference may give reasons that differ from the ones the candidate has indicated in the interview—and it is valuable to learn of them.)

11. **Would you rehire or work with this candidate again?** (I want to hear that this candidate was good to work with and that the manager would take the candidate back. If I'm

using a peer reference, I ask whether the peer would like to work with the candidate again.)

12. **What is the candidate's current or most recent salary?** (If salary is a go/no-go indicator for you, you may need to verify the candidate's stated current salary with the previous manager. Even if you've already asked the candidate, ask his or her previous manager to confirm. When making an offer, I don't determine salary based on the previous salary but rather on what value this candidate has to me, here, in this company.)

When I'm hiring a manager, I ask three additional questions, or variations thereof:

13. **How many people reported to this candidate?** (After asking this close-ended question, I follow up by asking if the people were easy or difficult to manage. I find out if the position I have open is something the candidate knows how to do, or will need to be taught. In other words, I find out how much managerial experience the candidate has. When I ask this question, I ask the reference to describe what easy-to-manage or hard-to-manage means to him or her. I once interviewed a management candidate who had only two direct reports. However, they were people whom the organization had declared to be "unmanageable," and my candidate had the job of helping them learn how to work with the rest of the technical group. Before managing those two people, he had managed a group of fifteen people, all of whom had sufficient social skills to work within the organization and sufficient technical skills to do the work. Such a candidate clearly had the needed managerial skills for the job.)

14. **How did the candidate conduct performance appraisals?** (If necessary, I prompt the reference, "Did the candidate conduct performance appraisals on an ongoing basis, once a year, or in some other way?" I want to find out whether the candidate did appraisals on time or late. I also want to know how the manager gathered data to write up performance appraisals.)

15. **How did the candidate handle "difficult" people?** (Anyone can be "difficult" at some point, for any number of reasons. I

want to know what kinds of people the management candidate had trouble dealing with, and how the candidate handled those people.)

Get your call to go through to each reference.

When you call another manager for a reference, you're probably trying to reach someone as busy as you are, someone who may not put returning your call at highest priority. When you ask for references, request cell phone numbers as well as work numbers, and ask for names and numbers of other people who may be able to help you get your call through to the reference. Ask the candidate to tell the reference your name in advance of your call, advising the reference that you will be calling within the next few days. As soon as you think you might make an offer, call the reference. If you're lucky, you'll speak with the reference when you call; if not, you can either leave your number for a call-back or you can make an appointment to speak to the reference at a later, mutually convenient time.

If you leave your number, make it easy for the reference to call you. For example, if you're calling a senior manager, such as a CEO or a Senior VP, ask his or her assistant to schedule a time for the executive to call you. Specify how much of the executive's time you expect to need—I usually like to have fifteen or twenty minutes—and explain that you will wait for the call at the appointed time. If you're hiring a senior-level staff member or a mid- or senior-level manager, ask for more time for the reference-check. The more senior the person, the more he or she will affect the work of other people. It's worth the time to check out these candidates.

Some references are prohibited from supplying any information about a candidate and will resist calling you back. If you think you are trying to get in touch with a less-than-enthusiastic reference, lay your cards on the table when you leave a message: "I'm checking references for Wally Walrus. Please call me back to provide a recommendation if he was excellent."[2]

Some hiring managers prefer to limit their reference-checks to fifteen minutes. If you are concerned that a reference won't spend more than fifteen minutes with you, choose a subset of your questions so that you can keep to the fifteen minutes. I recommend the following questions for a quick reference-check:

---

[2] Pierre Mornell advocates calling references when you expect they'll be unavailable so that you can leave a message saying, "Please call me back if the candidate was outstanding." See 45 Effective Ways for Hiring Smart: How to Predict Winners & Losers in the Incredibly Expensive People-Reading Game (Berkeley, Calif.: Ten Speed Press, 1998), p. 124.

1. Where and in what capacity did you work with the candidate?
2. How long did the candidate work for you?
3. If I were the candidate's manager, what advice would you have for me?
4. Would you hire or work with this candidate again?

If the reference is unavailable to be interviewed, ask the candidate for another reference.

Check references as completely as possible—even when the candidate has provided few, unreachable, or no references.

People change jobs, switch professions, move across the country, and even retire to remote Pacific islands—including the very people your candidate may name as references. If your candidate has no work experience whatsoever, has references who only speak a language that is foreign to you and your team of interviewers, or can't locate the people who could have served as references, you still have several options:

1. You can make an offer that keeps employment contingent upon the candidate's performance during a specified probationary period.
2. You can ask the candidate to provide the names of professional-society or networking acquaintances. For example, if the candidate has participated in a local networking group, ask for contact information for one or more of those colleagues as a reference.
3. You can ask for references from a non-profit activity (a school or religious organization, for example).

In the second and third cases, information collected in your reference-check may not be as reliable as you'd like, but at least you'll have some data to consider.

If a candidate is unable to provide an acceptable reference, ask for contact information for a pastor, a rabbi, or a teacher familiar with the candidate's history; as a last resort, offer the candidate the position on a probationary basis, citing a period of three months, during which time you can monitor the candidate in action as an employee. When you offer probationary employment, explain to the candidate that if he or she does not perform the job up to expected standards in those three months, you may choose to terminate the relationship.

Establish rapport during a reference-check.

During a telephone reference-check, you need to establish rapport quickly and learn as much as you can in your allotted time. To show the reference you respect his or her time, start the conversation quickly and then listen carefully to the answers.

Start the conversation quickly.

Introduce yourself, confirm that this is still a good time to talk, and ask your first question. If applicable, explain that you'll be taking notes during the conversation and may need a few seconds between questions to record the answers, and then jump right in to ask your questions.

If the reference indicates that he or she no longer has time available, make a new appointment. If a reference postpones your conversation more than once, ask the candidate for another reference.

Listen carefully to the answers.

When you conduct a reference-check over the telephone, you cannot take clues from body language and so you must be careful to listen for the intended meaning in *ambiguous* answers. Clarify the real intent behind the response while you still have the reference on the phone. Although the following comments are somewhat humorous when analyzed, they point out the danger of accepting words at face value:

- *"I can assure you that no person would be better for the job."*[3] (Does the reference mean that hiring *no one* would be better than hiring the candidate? Is the candidate *better than nobody*? Does the person making the assurance really intend to convey a thumb-up vote?)
- *"He has a wicked sense of humor."* (By "wicked," does this reference mean that the candidate is sarcastic, dour, insulting, caustic, and severe, or does the comment mean that the candidate is truly clever and sharp-witted?)

---

[3] This example of an ambiguous recommendation, directly quoted by Robert W. Wendover [*Smart Hiring: The Complete Guide to Finding and Hiring the Best Employees,* 2nd ed. (Naperville, Ill.: Sourcebooks, 1998), p. 145], comes from a set of responses put forth by Robert Thornton in *The Lexicon of Intentionally Ambiguous Recommendations* (Minnetonka, Minn.: Meadowbrook Press, 1988).

Also listen for *confirming* answers. If you ask a question about how quickly the candidate learned about a product line, and you hear "As fast as everyone else," ask, "How fast was that?" If the initial response is "Very fast," then ask, "How did you know?" If the response is "Slower than expected," ask, "How did you know?" and "What did you expect?"

I don't recommend using e-mail for reference-checks. People are much less likely to take the time to answer you, and they rarely answer in a timely fashion.

Verify employment, salary, and education claims.

I usually seek confirmation of a candidate's employment history back through the most recent two or three employers unless there is something specific in the candidate's prior work experience that I feel the need to check. I personally do include the salary check because every organization I've hired for has viewed salary verification as important, but you can take your cue from your own company's policy. My rule of thumb is, if the salary the candidate reported doesn't strike me as probable or doesn't match the salary the organization reported, I ask the candidate to account for the difference. If he or she can provide a reasonable explanation for the difference, fine; if not, that's not fine.

Educational degrees are not a reliable predictor of work success, and so I don't usually bother verifying claims of academic accomplishment. However, in an organization in which degrees are important, such as in a technical lab or in a management-consulting company, verifying educational claims can be very important.

I decide how important verifying a claim about education or salary is by applying the following test: If some years from now, word leaked out that the individual's claim regarding salary or education was false, what would be the effect? If the answer is that there would be no adverse effect, then I don't bother verifying the claim. If the consequences of learning the truth at a later time will be serious, then I verify the claim.

Incorporate other checks that are required by your organization in the reference-check.

Some organizations require that a background check be run on potential employees. If your organization performs work for the government, even if the work is not classified, the background check probably is a requirement for employment, regardless of whether for a technical position or other. Your company may also require other tests or assessments. For example, some companies use handwriting analysis, personality assessments, and physical tests to verify that a candidate does not abuse alcohol or drugs.

Sometimes, a personality assessment can be used to predict how well a person will fit into a team. However, I've found that the people who use personality assessments often attempt to fit people into rigid molds, molds that do not allow the various components of the candidate's personality to show. Personality tests can be used by employees themselves to help them understand their own preferences for how they work; I do not like to use personality tests as a hiring tool. If you do use personality assessments, make sure you're not using them to discriminate against candidates.

If your organization requires checks, tests, or assessments of a candidate's physical condition, make sure your personnel or HR group takes the responsibility for those checks. When I check references, I ignore physical tests and assessments, unless there is something that would prevent the candidate from satisfying the physical requirements of the open job. For technical people, unless the issue is drug or alcohol abuse, physical testing is rarely required.

Take action to uncover the truth if you find discrepancies.

Although this happens infrequently, I sometimes find a discrepancy between what a reference tells me is fact and the information on a candidate's résumé—usually pertaining to schooling or coursework, responsibility at work, or a title. When I uncover a discrepancy, I review both sides of the story and assess how relevant the discrepancy is to the candidate's ability to perform the job. Factors I consider are discussed below:

- *Did the candidate claim a specific title, such as Senior Engineer, when the company listed a different title, such as Engineer?* If a candidate's title claim is in question and you have access to his or her application for employment with the previous company, the application may explain the discrepancy. If not, ask the candidate and, if possible, the candidate's manager for clarification. Discrepancies can result from the practice of some candidates to use a standardized title on the résumé so as to match industry terminology. Other discrepancies may arise with candidates who have been told they were in line for a promotion, or with those who may even have been given the promotion, but the paperwork isn't completed before the candidate leaves the position. I generally don't care what title an applicant claims, as long as the claim matches the candidate's experience and salary level and the candidate is willing to discuss the salary and title I have to offer.

Management titles are not safe from discrepancies. What one organization calls a director, another calls a group manager, and yet another calls a vice president.   The candidate may have listed a title that he or she thinks reflects more accurately what the industry perceives the title to be. Ask the candidate or the reference what the company-given title was, and what that title meant.

- *Did the candidate lie about graduating from college?* Some people are very insecure about appearing as if they don't have a degree, and may stretch the truth on a résumé, figuring that most companies don't check college records. Because I usually consider life experience and on-the-job training to count as much as a diploma, I do ask the candidate to explain the claim. Based on the response, I then determine whether or not to pursue the candidate. In fact, I *am* troubled when I discover that a candidate has lied *about anything* on his or her résumé, but I view an academic claim in terms of its relevance to the job, asking myself, "How important is the claim to this job?"

- *Did the candidate claim to have completed certain courses, but doesn't appear to have the associated knowledge?* This kind of discrepancy is important to me because it points to the heart of the reference-check: Can this candidate perform the job well?

- *Does the candidate's stated salary match the salary reported by the company?* If I encounter this type of discrepancy, I ask about base pay; if the numbers are significantly different, I go back to the candidate to find out what he or she included in the salary statement. Sometimes, candidates include a bonus as if it were salary, but their company states salaries independent of bonuses. Some companies include the dollar value of certain benefits when citing salary, but their employees report salary without factoring in benefits. It is important to determine the cause of the discrepancy.

When you discover a discrepancy between what the candidate has reported and what the references say, dig deeper. Ask the references about specific projects the candidate discussed in his or her interview. If you find discrepancies, review the information brought forth in the interview, and determine the cause of the discrepancy. If the discrepancy occurred because the applicant lied on his or her résumé or in the interview about something that matters, you can decide what action to take. If the discrepancy stems from one small source—such as a title discrepancy that results from a difference in terminology—and everything else checks out, I generally make an offer. I'm very concerned, however, if I discover a *pattern* of discrepancies, combining, for example, title inflation, a claim of inflated responsibility on a project, and an attempt to take credit for other people's work. If I find a pattern of discrepancies, I drop the candidate, and I let any affiliated recruiter know about the lies.

POINTS TO REMEMBER

- Always check references.
- Build your reference question list in advance of the reference-check, and note the answers you receive when you call references.
- Word questions so as to elicit answers that are confirming and unambiguous and that will enable you to determine both that the candidate can perform the job and how well the candidate can perform that job.

Creating, Timing, and Extending an Offer

Hiring Manager: "Hi, is this Damon R. Candidate?"

Candidate: "Yes, this is Damon speaking."

Hiring Manager: "This is Joseph Manager, from the performance group you interviewed with yesterday. I'd like to make you an offer for a job as a Performance Software Engineer. Here are the particulars: We can offer a salary of forty thousand dollars, thirty thousand shares of stock options that vest evenly over three years, and our standard benefits."

Candidate: "Oh. Thank you, but I'm afraid I cannot accept an offer of less than fifty thousand dollars in salary."

Oops. When you make an offer, you want to know that salary and other conditions are in the acceptable range for a candidate. To do that, you need to know as much as possible about what the candidate wants from a job, and what you're willing to provide.

If you asked the candidate about his or her salary requirement in the initial phone-screen, you may have heard a good indication of the desired salary or, at the least, the salary range.[1] However, many candidates state that salary is "open to negotiation," which doesn't help you when you want to prepare an offer. Of course, if you're working with an external recruiter, you can ask the recruiter to tell you what the candidate is seeking, but an effective way to prevent wasting time on a candidate who

---

[1] For practical advice on when to talk about salary, see Lou Adler, *Hire with Your Head: Using Power Hiring to Build Great Companies*, 2nd ed. (Hoboken, N.J.: John Wiley & Sons, 2002), pp. 216-19.

seeks a salary far out of your range is to tell the candidate the range at the time of the phone-screen or interview, verifying then that the candidate can accept the salary if offered the job.

It isn't as simple as that, however. When candidates evaluate offers, they may be willing to take a lower salary if some other benefit sweetens the pot. For example, a candidate may accept a lower salary if offered stock, a performance-based bonus, a chance to work in a particular field or organization, easy access to company-sponsored day care, additional vacation days, a company-leased car, or any of numerous other benefits. Before you make an offer, make sure you have discussed what's important to the candidate—either with the candidate during the phone-screen or interviews, or with the internal or external recruiter, as applicable.

When you consider salary, decide what you think the candidate is worth *to your organization and in the position you have to offer*. There's no point in developing an offer if the candidate wants more money than the position is worth.

In a strong economy, make your offer soon after the last interview.

When the economy is good and the job market competitive, you may want to extend your offer to a candidate on the same day as his or her last interview. Make same-day offers contingent on a background check and positive reference-checks, of course, and prepare them by drawing up several versions for which you can get all necessary written approval *in advance*. Depending on your company's policy, you may need to get approval from an officer of the corporation or from an authorized representative from the Finance, Legal, or Human Resources Department. Whatever your company's policy, don't let the right candidate languish while you run around trying to get approval for your offer.

Making a same-day offer requires flexibility from all people concerned:

- You may need to offer an "unusual" package, including one or a combination of the benefits and perquisites discussed in greater detail later in this chapter.
- You may need to change the salary offered if you want the candidate to accept it on the same day as his or her last interview.
- You may need to secure additional signing authority for yourself or be guaranteed of access to people who, other than yourself, can sign offer letters late on any given day.

Same-day offers can test the flexibility of an organization. If you think you will want to make such an offer, discuss the procedure well in advance with your management and other relevant stakeholders in your Finance, Legal, and HR Departments.

For every offer, review all components before presenting it to a candidate.

Because a compelling offer consists of more than just salary, ask the candidate during the interview process what other components would be required to make an offer attractive enough to accept.[2] If the candidate has told you he or she is currently considering another offer (or if you merely suspect this to be the case), it is quite acceptable for you to say, "I understand that you may be weighing several offers. Can you tell me what criteria could be key to your decision?" If you uncover something specific the candidate wants, such as an assurance of no weekend work or of a position on the 4 P.M. to 11 P.M. shift, and you are willing to guarantee it as a condition of employment, add that detail to your offer.

Although not everything is negotiable even in the most flexible of companies, there are some areas you may be able to tailor to the candidate's requirement. Following are areas that your company may consider negotiable:

- job title
- work assignments
- vacation time
- salary
- stock options
- tuition reimbursement
- professional training
- association memberships

The question "What will it take for you to say yes to an offer?" sometimes surprises candidates, so they may not have an answer for you immediately. Give the candidate a day or so to consider the question while you check references. Then, talk again with the candidate before you put together the offer. Listen carefully to what the candidate says—he or she is giving you very specific information as to what's important.

When you make an offer within twenty-four hours of the candidate's last interview or as a same-day offer, you communicate the following messages:

---

[2] Why it is important to discuss all components of an offer with the candidate, *before* making the offer, is handled especially well in Adler, op. cit., p. 219.

- You value the candidate and want your intention to hire known.
- You are willing to expend considerable effort to make the offer attractive.

Beware of making promises you may not be able to keep.

I've had people reject what seemed to me to be perfectly reasonable offers, but most often, people shy away from offers that include promises of perquisites and benefits that the candidate will receive at some point in the future, well after he or she has started the new job. The reason candidates view promises of future perks and benefits with skepticism is that many candidates have been burned by promises that were not kept. Understandably, such people are concerned that once they start work, the hiring manager will not have the authority to fulfill the promise. Promises that candidates view with the most skepticism are promises of a promotion or of a bonus within an agreed-upon number of months after the candidate has started work and promises of special-order equipment or flexibility in office location.

If you or any members of your team are inclined to include any of these four promises in an offer to a candidate, you should take precautionary steps to head off a skeptical reaction from the candidate. In the following paragraphs, the discussion of promises that could create cynicism or bad feelings toward your organization is paired with possible resolutions. By presenting promises intelligently, you can reduce the probability of the offer being rejected.

Promise of a promotion

If you make an offer that promises that a candidate will receive a promotion—and an associated salary increase—at the end of, say, six months, some negative event may occur during the specified period that will preclude the promotion and raise. For example, the company's quarterly earnings may fall below predictions, causing the Board to order a wage freeze into effect; or, the company may merge with another company, resulting in massive layoffs and postponed promotions among the survivors; or, the company's management may decree that the candidate does not merit the promotion. Whatever the reason the promise is not kept, you're stuck with an unhappy employee.

Promising an unconditional promotion is not just risky; it is stupid. Circumstances within the company can change; the employee may not perform up to expectations; the economy may tank. If you decide to promise a promotion, be specific with your candidate about the conditions that need to occur within what time period. Describe what performance is expected of the candidate once employment begins, but also describe what performance is expected of the company and the marketplace in order for the promotion to be viable.

Because it is impossible to predict the future and almost as difficult to accurately detail performance expectations, I do not recommend promising a promotion as part of a job offer. Instead, promise the candidate that a formal, performance review will be held at the end of three months (or six months or whatever time frame you prefer). Providing a performance evaluation doesn't require the organization to be profitable, a merger to be ignored, or the economy to be strong—thereby making your promise one that can be kept. And although you can, of course, give performance feedback to your employee at any time, promising a specific time for the review shows the candidate that, as an employee, he or she will be treated in a professional manner by a well-organized management.

Promise of a raise or a bonus

If you ever were hired to work at a company when business was flourishing, and you bought into the idea of accepting a lower starting salary because you were promised you'd receive a raise or bonus within a specified period of time, and then your company went through a slow period, you undoubtedly missed out on getting the promised raise or bonus. You are not alone. The sad fact is that it is not uncommon for companies to offer reduced base salaries because they plan on giving bonuses later, and then to eliminate issuing bonuses because profits don't match expectations.

If, as the hiring manager, you want to pay a lower base salary and make a bonus part of the job offer, make sure you confirm that your organization does anticipate paying out bonuses, and clarify under what conditions bonuses will be allocated and to whom. Make sure your candidates know that the bonus is contingent on profits. Discuss the salary structure with your HR group and your management to make sure you can adequately pay staff salaries plus bonuses.

Sometimes, a bonus is contingent on some objective. If the objective is something the employee has control over and can accomplish by working alone, then a bonus can be a reasonable component of an offer. However, since I know of no technical work that can be accomplished by one person working alone that is worth a bonus, I view earning such a bonus as unachievable for most, if not all, technical staff members.

If the bonus is contingent not on profits or productivity, but rather on some other non-revenue criterion, then your organization may be postponing part of the employee's salary in favor of managing its own cash flow better. I regard this practice as unethical and therefore don't work for companies that do it, but if your company follows this practice, make sure you can explain to the candidate why making his or her bonus contingent on whatever non-revenue criterion your company measures is acceptable.

If you are involved with hiring technical staff members, you'll undoubtedly run into candidates who want you to clarify how they can meet corporate goals in order to qualify for a bonus. This is a reasonable request as technical people often work like crazy to meet deadlines and finish a product, but their organization has some other problem that prevents enough revenue to be generated for bonuses. Middle-level managers on the technical staff may be able to meet corporate goals with their work, but it is rare for a first-line manager or technical person to meet goals set by the corporation. As a result, many technical people with more than a few years of experience are leery of promises of bonuses. If your organization makes a bonus part of an offer, be very clear on how your technical staff candidate can be assured of collecting it.

The following dialogue illustrates the frustration so many technical staff members experience:

Manager: "Richard, I'm extremely pleased to have you on my technical staff and have put through a small raise to coincide with this first evaluation."

New Hire: "Thank you, but can you tell me when I'll receive my bonus?"

Manager: "Oh, that's on hold for now. All bonuses are frozen this year."

New Hire: "Did I meet my goals?"

Manager: "Yes."

New Hire: "Then I should be receiving a bonus now. In my offer letter, you indicated that if met my goals, I'd receive a bonus. Is the company reneging on my offer letter?"

To save yourself and your candidates unnecessary frustration, offer something else that a technical person is likely to want, to be awarded on the basis of his or her performance of great work. A promise to send the candidate to a conference or training course, or a promise to add vacation days to his or her benefits package, is more likely to convince the candidate to accept your offer than is the promise of a bonus.

Equipment or a flexible work location

Sometimes, managers promise such signing incentives as a company-provisioned laptop, a cell phone, a cable modem hookup, reimbursement for home-office equipment, or the possibility of working from home. If you are tempted to make these kinds of promises, check with your management first to make sure the company is willing to take on the expense.

In general, any promises based on a new hire's joining the staff first and receiving the reward later are more likely to convince the candidate to say no to a job offer, rather than yes. One technique that may make a promise more enticing is to incorporate it in the offer letter, because the candidate is more likely to believe a promise in an offer letter. However, most candidates I have worked with are not inclined to believe a promise, and may decide your entire offer—not just the promise—is bogus.

Make the offer easy to accept by including perks and benefits you *can* deliver.

Salary alone may not be enough to convince a candidate to accept your offer, although an inadequate salary may prompt him or her to reject the offer regardless of what other enticements have been included. To put forth your company's best offer, you may want to use some or all of the following ideas, keeping in mind that what makes an offer attractive depends on what a candidate truly wants, as well as how risk-averse he or she is.

Offering perks or benefits that deal with expenses instead of salary can make your job more difficult. If you offer expense-account perks, make sure you, your management, and the number-crunchers in the Finance or Bookkeeping Department understand how to manage your budget, so the perk isn't denied when an employee tries to use it.

Company stock

If your company is able to add stock to the salary offer, this can provide a strong incentive to the candidate to join your group.

In a start-up organization with more promise of growth than actual operating capital, stock offered in conjunction with a lower initial salary (instead of *in addition* to it) may appeal sufficiently to certain entrepreneurial candidates to get them to say yes to the job offer. For others, you will need to be able to offer stock in addition to salary, frequently a realistic option in well-established companies or in start-ups with adequate funding.

In a financially sound organization, you may be able to make the offer even more appealing—by awarding stock either with a short vesting period (say, three years or less) or to be issued at performance-evaluation times. Such incentives show candidates that there is benefit in working for your company for a period of years, without forcing them to have to worry about the company's cash flow—a concern they would have with a smaller salary and more stock.

When you make stock part of an offer, make sure the candidate understands there is risk—stock may decrease in value rather than increase. This possibility was brought painfully to my attention when a colleague, who had accepted stock as part of his compensation when he first went to work for an established, closely held consulting firm, decided to leave the company after ten years and attempted to sell his stock. He discovered that although the company valued the stock upon its issuance to him at $70 per share, because he had no buy-back agreement with the company, the company had no obligation to buy back the stock. Eventually, he was able to sell his stock—for $20 per share.

If the management of a company has no intention of ever taking it public or of merging with a profitable company or of accepting an acquisition offer from a cash-rich company, paying people with stock in lieu of a higher salary strikes me as ethically untenable. If there is never likely to be a general market for the stock and the company has no buy-back agreement, that stock is useless as a part of the compensation package.

Vacation time

If you're having trouble meeting a candidate's salary requirement, consider whether you can offer additional paid vacation time as a benefit that will add value, perhaps even enough to equal the difference between the salary you are offering and the salary the candidate desires. By authorizing more paid vacation time, you will, in effect, be paying the candidate more money for each hour worked. By adjusting vacation time and salary up and down as if they were weights on two sides of a scale, you may be able to stay within your budget while giving the candidate more of what he or she values. This can work to your mutual benefit in several ways:

- If a candidate wants a higher salary but also wants more vacation time, you may be able to grant the higher salary on the condition that he or she will take an agreed-upon minimum number of added vacation days as time without pay. With this arrangement, the employee can take some time off as he or she wants, without costing you the full amount of the higher salary. Another advantage of this arrangement is that it does not jeopardize salary parity within the rest of the group.
- If a candidate wants a higher salary, but you cannot grant it because you must maintain salary parity within the department, you may be able to offer him or her a package to include an increased number of paid vacation and personal days, paid floating holidays, and the like. By paying for more time off, you are, in effect, paying a higher salary.

Conferences, training, and other career-enhancing opportunities

If you've offered a fair but not outstanding salary, you've probably met the candidate's base-pay requirements. If you add career-enhancing opportunities to the package—such as company-paid conference attendance, training classes, or a book-purchase allowance—you can preserve your company's salary outlay while increasing the value of your offer to the candidate. You can take this approach if you need to keep within a salary budget but have a separate pot for non-salary compensation.

For example, you could offer a candidate the opportunity to attend conferences, up to a total expense of, say, $3,000 per year. (In truth, $3,000 barely covers registration, travel, hotel, and per-diem expenses for one conference these days, but I'll stick with the number for argument's sake.) If the individual wants to stretch the allocated money, he or she has numerous options, possibly choosing to take the cheapest flights; to stay in a budget hotel; to present a paper in exchange, perhaps, for complimentary enrollment; or to attend only local conferences. Depending on an individual's preference, he or she either could attend numerous conferences within a year and still stay within the yearly budget or could blow the whole amount at one fancy boondoggle in some exotic resort. The fact that the candidate can choose how he or she uses the money can be viewed as an additional benefit in itself.

There are other benefits that you can offer in place of a higher salary. For example, if you regularly bring training in-house or if your company offers tuition reimbursement for accredited advanced education, explain those benefits to the candidate, and be prepared to quote the associated dollar value if asked. If you don't bring training in-house or reimburse for college credits, consider offering the candidate a training budget to use as he or she selects, to a maximum amount.

A book-purchase allowance is also a valuable benefit that you may be able to offer in addition to salary. Make sure you specify who owns the books, however, so that at the end of the candidate's employment, you don't have a misunderstanding. If the employee will own the books, then selection and ownership of the books both are perks. If the company will own and retain the books, the benefit to the employee comes from being the one who chooses the books and the one who can read and reference the books during his or her period of employment. If you offer a book-purchase allowance but the individual can't take the books away at the end of his or her employment, you're not adding dollar value to the employee's salary, but you are adding value.

If you will offer career-enhancing opportunities as an incentive, I suggest you define which opportunities you can offer at which job levels. That way, you will be able to treat people fairly, and avoid the possibility of inadvertently compensating one employee with an incentive reserved for a different job level.

Choice work assignments

When you look for a candidate whose skill and experience satisfies the requirements detailed in a particular job description, and you encounter someone whose expertise is in a related but separate area, you may want to consider the candidate anyway. For example, you may find that the candidate wants to perform additional assignments or to change his or her line of work. Over the years, I've successfully filled positions with testers who wanted to improve their automation skills, developers who wanted to improve their design skills, and technical support representatives who wanted to be able to debug software. To make it feasible for these people to perform the work they wanted, I helped set up in-house, apprentice-type training sessions for these employees so that they could learn enough to start performing the work they wanted. In every one of these cases, my company and I benefited by having an enthusiastic staff member who was eager to work at peak capacity.

Before you agree to grant a candidate his or her choice, confirm with the candidate's future coworkers and managers that they are willing to provide all necessary training. Some managers will not agree to the hiring because they fear that the new hire may be just using the position to get a foothold in the company, and will jump ship to the area for which he or she has previous experience as soon as a position there opens up. I personally am willing to take this chance, believing that a developer who gets trained to think as a tester, for example, will be a better developer when he or she returns to development. One way to provide a bit of insurance against someone jumping ship before the training investment has been recouped is to make any transfer within the company conditional on the individual's completion of a minimum number of months of work (say, six or nine months) and on written approval of the transfer from all affected managers. Whether the conditions are stated as a request or as a requirement, the decision should be based first on what's best for the company, and second on what's right for the employee.

Personal attention

There are other ways to attract candidates than piling on perks and benefits. In fact, many people join an organization because they like the people with whom they will work. You are one facet of the people picture the candidate sees prior to joining your company, so take care to present yourself—both as a professional and as a person—in the best light. Be responsive to any questions or requests that the candidate might have during whatever period he or she is given to consider the offer. Ask whether the candidate would like additional time to talk with members of the prospective team. By understanding that most candidates don't decide to accept or reject a job offer solely on the basis of money, you can strengthen your offer by providing the candidate with personalized attention, non-monetary benefits, and ready access to the people already on staff.

Learn the reasons behind a candidate's rejection of your offer.

If a candidate rejects your offer, it makes good business sense to learn why. Although receiving a rejected offer clearly is disappointing, the decision can provide you with an important opportunity to gather information. Ask the candidate to share his or her reasons for the decision, saying, "I'm disappointed you decided not to join us. I would appreciate it if you could share your reasoning with me."

It is a fact that candidates reject offers for reasons other than salary: It could be that the individual has taken another position. Perhaps he or she has opted to work in a different environment or with a different team of people. Until you learn why the candidate rejected your offer, you won't know how to proceed.

When the reason is salary, salary, salary, rethink the offer.

If salary is cited as the reason for the rejection, the candidate may believe that his or her potential value to your company is higher than you think it is. Assuming you want the candidate on board but have little money to add to the pot, you can respond in various ways: You can increase your offer either by raising the salary by whatever amount you have left or by adding perks and benefits to the base salary; you can keep the offer as tendered but hold it open for an extended period of time so that the candidate can weigh it against other offers; or you can explain that you cannot improve the salary and that you will need to move on to other candidates. If you hold the offer open, make it clear that you understand that the candidate wants to wait to see what happens with the other offers, and ask him or her to get back to you once all options have been evaluated, or, if you prefer, by a specific date.

I generally don't suggest increasing the salary once an offer has been presented, in part because I don't want to enter into a bidding war with other potential employers or with the candidate's current employer. Instead of getting into a bidding war that is likely to leave both sides asking, "What amount—high or low—could I really have gotten away with?" I prefer to say to a candidate, "This is my best offer. I'd really like you to work here for these reasons," and then I list the reasons. I want the candidate to know that I do value him or her as a potential employee; and sharing the reasons should give me more credibility as a responsible and professional hiring manager than I could earn by granting a gratuitous salary increase.

Sometimes, improving the salary does constitute the most reasonable approach. If this particular candidate has the skills, qualities, and preferences you want, and you're not willing to take the risk of waiting for another potential candidate to cross the threshold, reevaluate your offer. A few words of advice: If you find yourself reevaluating offers frequently, ask yourself what's really happening: Are you stuck using salary ranges that don't match what people should reasonably expect, salaries readily offered by your competitors? Are you undervaluing a candidate's worth to your organization? Are you looking for candidates with more experience than you're willing to pay for?

If your organization's salary ranges don't make sense compared to a candidate's expectations, first do some detective work. Ask around your own and other organizations to determine whether your peers—other hiring managers—have the same problem. If so, review the on-line salary surveys, such as at salary.com, or buy a salary survey from a provider that surveys technical employees—and compare your salary ranges against comparable organizations in similar geographic locations. If you still have a problem with salary ranges and parity, take a look at everyone's salary—it may be time for an adjustment for everyone in your group, to bring them up to par with the external world.

Don't be cheap with technical staff salaries. If you're running into general salary problems, work with your management and the HR people to resolve them. This is an area where your HR staff probably has the expertise you're missing. Use HR to make the salary ranges work for you.

If you're looking for someone with specific, hard-to-find skills, and you need this person to start quickly, be prepared to pay more. For example, if you're looking for someone as scarce as a release engineer, then be prepared to pay more than you may have initially allocated. Hard-to-find skills make an employee more valuable in the marketplace, even if the work the person will do for you isn't worth more.

If your salary ranges are reasonable but candidates reject your offers, citing salary as the reason for the rejection, review how you're evaluating the candidate's worth to your organization. Why do candidates think they are worth more to you than you think they are worth? Don't make the mistake of keeping offering salaries down because you don't want any members of your staff to make more money than you make. Senior-level, highly skilled technical people often do make more money than hiring managers, so evaluate the candidate fairly, looking at his or her worth to your organization; review what the candidate's peers earn; and ignore how the salary looks in the hierarchy.

Know when it's okay to offer a job to an over-qualified candidate.

Sometimes, you will interview an experienced candidate for a junior-level position. This happens most commonly when the economy is slow or when a person who has been out of the job market for a lengthy period of time wants to reenter. Be clear with yourself about what you're willing to pay for the position, and then find the most appropriate person at the salary the company is willing to pay. Just because a candidate has more than enough experience for a position doesn't mean you have to pay that candidate more than the job is worth. Keep in mind that a senior-level, experienced person who is offered a junior-level job and a lower salary does not have to take the job, but many will opt to accept because it is the best option open. Especially in a slow economy, a person who has been out of work for six months or more is likely to appreciate having a job, even if the pay is not what he or she received before. Do not use a candidate's experience level or recent bout with unemployment as a reason for either increasing or decreasing the salary.

While a person who has been out of work may appreciate having a job, underpaying an employee is not a long-term retention strategy. When you pay a salary that is significantly lower than the realistic market value, you can expect employees to leave for greener pastures as soon as the market improves. They'll leave, that is, unless you give raises when the market improves. Underpaying employees is a false economy because it results in turnover, increasing your cost to hire. Replacing someone you hired on the cheap can cost you many times what you saved initially in salary because you will spend many times that amount later in training a succession of people who use the position as a jumping-off place. The cost of such turnover was estimated at $150,000 more than a decade ago.[3] Certainly, turnover is not less expensive today.

So, how do you protect your investment when you hire a person who is over-qualified and underpaid for the job? One way to manage the risk is to be totally honest with the senior-level person when you offer him or her the job: Ask individuals who accept such a position to tell you of their job search before they start looking for a new job. Let them know that, if possible, you will increase their responsibilities and salary at such time as you are able, but that this is not a guarantee or a promise. If the candidate is ready to take your position at your salary, offer it.

Close the offer.

When I make an offer to a candidate, I always ask the candidate to give me a decision within five business days. Specifically, if I get my offer letter into a candidate's hands on a Wednesday, I want to know his or her answer on the following Wednesday, at the latest.   My work as hiring manager is almost done when the offer letter arrives at the candidate's

[3] See Tom DeMarco and Timothy Lister, *Peopleware: Productive Projects and Teams*, 2nd ed. (New York: Dorset House Publishing, 1999), p. 207.

home or office, but my schedule leading up to that moment requires a high degree of organization and a modicum of luck. For example, to get my offer letter to a candidate on a Wednesday, I would have scheduled all my interviewers to complete the interviewing process by Monday afternoon, and I would have needed to check the candidate's references that night and on Tuesday morning (this is where luck can be needed, as I'll want to be able to reach and talk with all references during one evening and the following morning—not always doable). On Tuesday afternoon, I would have had to develop the offer, get approval from management, HR, or others, as necessary, and then telephone the candidate to present the offer verbally. Assuming that the candidate wants to proceed, I would then have had to prepare the written offer and send it via overnight courier for delivery on Wednesday. In my offer letter, I ask for a decision on or before the following Wednesday. The two days leading up to putting the offer letter in the candidate's hands are hectic, but the tight schedule is an important component in closing the offer.

Limiting the amount of time the candidate has to consider my offer is part of my hiring strategy. I have observed that, in general, candidates who need more than one week to decide are not evaluating my company on its merits; they're comparing offers. I want the candidate to be motivated to work for me, in my organization. I don't want the candidate to take a job simply because it is the best monetary offer.

I make an exception to limiting the candidate's response time to one week if the period during which the candidate was interviewed and during which his or her references were checked stretched out over days, weeks, or even months. If you take a leisurely approach to hiring, you cannot expect the candidate to respond quickly. You've already trained the candidate that you move slowly, and the candidate will reasonably assume that he or she can move slowly as well.

Before closing the offer, you'll need to set the candidate's start date. I recommend limiting the time you're willing to wait for the candidate to start work to three weeks. This block of time allows candidates to give two weeks' notice at their current company and then to take one week off before starting with you. If a candidate requests more than three weeks, I listen to his or her reasons for needing extra time, and then I evaluate whether the reasons seem legitimate. I become concerned if the reasons are unconvincing, having encountered a few people over the years who used the extra time to continue interviewing and to compare offers, and then called my office the day before they were supposed to start to say they weren't starting. Although employing such a person probably would not have worked out for the long term, the experience is unpleasant and disruptive, in most cases leaving you without enough time to begin interviewing again. If this happens, you may be forced either to hire someone who isn't fully qualified, or to shuffle work assignments among existing staff members because you've run out of time.

Sometimes, a candidate has a perfectly legitimate reason for asking to postpone the start date or for requesting time off during a specific period in the future. I usually agree to the candidate's schedule if he or she has made it known to me at the beginning of the interview period or during my verbal offer. A candidate who asks for time to take a previously scheduled vacation, or who requests time to decompress after a particularly grueling assignment, is likely to be a better employee if he or she is granted the time. If such a candidate will make the commitment to your company by becoming an employee before leaving for the time off, I see no problem. If the individual won't sign on as an employee, you face a real risk that he or she won't ever start.

Use a standard offer letter.

When you're ready to make an offer, use your organization's standard offer letter. If you don't have a standard offer letter, draft one in advance for review by your corporate lawyer and your management. Always write an offer letter on company stationery because the offer commits the company, not you personally. In addition, keep in mind that an offer letter is a legal document made by you *on behalf of your company*, so make sure it is a commitment you want to honor.

In the offer letter, make sure you include the date of the offer; the job title for the position to be filled; the salary; any stock, vacation, or benefits in addition to what the company's basic policy stipulates; who the candidate's manager will be; the start date; and the reply date.[4] Let the candidate know that if he or she doesn't respond by the reply date, the offer is no longer valid and may be withdrawn.

An Offer Letter Template that you can customize for your own use appears on the following page. If you do use it—either as is or as the basis for your letter—make sure your corporate lawyer reviews it and then approves your version.

---

[4] One caveat: Never cite *annual salary* in an offer letter. Instead, state salary in terms of *one pay period*, and specify the number of pay periods per year. You do not want a candidate to view the offer as being for a job having a minimum of one year's duration and, consequently, to expect to be paid for the full year regardless of the actual term of employment.

Offer Letter Template

<[date]>

<[candidate name]>
<[candidate street address]>
<[candidate city, state, and zip code address]>

Dear <[candidate name]>:

I am pleased to offer you the position of <[job title]> *with* <[company name]>, *located at* <[location description]>, *reporting to* <[manager name and title]>.

1. *Your responsibilities will be those outlined in the enclosed job description and described to you during your discussions with me.*

2. *You will be compensated with a* <[weekly/biweekly/monthly]> *salary in the amount of* <[the salary]>. *Other compensation shall consist of* <[list of additional benefits/stock/perks]>.

3. *The Company has the following* <[number]> *pre-employment requirements:* <[physical examination/review of documents]>, *which will need to be satisfied prior to employment.*

4. *You are considered an "at will" employee. This means that we can terminate your employment with or without cause, and with or without notice, at any time, at the option of either* <[company name]> *or yourself, except as otherwise provided by law. Additionally, because you do not have an employment contract with us, you can terminate your employment with or without notice at any time.*

5. *Our offer to hire you is contingent upon your submission of satis-factory proof of your identity and your legal authorization to work in* <[country name]>. *If you fail to submit this proof,* <[federal/state/ local]> *law prohibits us from hiring you.* (Check whether you need this clause if you work outside the United States.)

6. *I hope you can begin work on* <[day, date, time]> *at* <[position location>].

If you agree with and accept the terms of this offer of employment, please sign below and return this letter to our office on or before <[day, date]>. *I look forward to hearing from you and to having you join us.*

Sincerely,

<[your name]>   <[candidate name (signature)]>   *Date signed:* <[date]>
<[your title]>   <[candidate name (printed)]>

Extend the offer.

To extend the offer, you can read the offer letter aloud to the candidate or you can communicate a summarized version of each of the points. It is important for you as the hiring manager to extend the offer yourself, first by telephone if possible, and then by letter. If calling the candidate is difficult to do because of your or the candidate's travel or work schedule, or for some physical reason, then have someone else verbally extend the offer. However, if you don't make the offer yourself, ask whoever represents you to explain why you are not personally making the offer.

POINTS TO REMEMBER

- Start learning early in the selection process what it would take for this candidate to say yes to your offer.
- Avoid offer-rejection triggers.
- Give candidates one week to make a decision once they have received the offer in writing.
- If you're the hiring manager, extend the offer personally.

Part 5

Making the Most of Hiring Opportunities to Control Uncertainty and Risk

Even if you follow all the advice in this book and you do everything just right, your job—hiring technical people—may still bring you face-to-face with moments of uncertainty and risk. The good news is that such moments also provide you with opportunity.

One of the biggest opportunities arises the moment your new employee shows up for the first day of work. Ensuring that a new hire has a great first day should be every hiring manager's objective, but you, like most hiring managers, undoubtedly have many items on your plate. If you think you cannot be available, think back to your first day at your company. Was it a great experience, or was it an experience you'd rather not remember? Whatever your personal experience, Chapter 13, "Creating a Great First Day," will provide you with ways to use opportunity to ensure that your new employee has a positive experience.

Hiring a manager may be the riskiest problem in all of hiring. Chapter 14, "Hiring Technical Managers," is designed to help you analyze the manager's job. In the chapter, I let you in on a well-guarded secret: Once you've analyzed the manager's job, the rest is easy.

In the final chapter of the book, I discuss how to handle the problem no hiring manager wants to have. Chapter 15, "Moving Forward," gives tips and techniques to use when you feel you have done everything you can think of and you still have no one to hire for the job.

13

Creating a Great First Day

"I showed up on my first morning at 8:30, just as I was supposed to. For the first hour, I sat in the lobby because I had no office. The facilities manager finally took me into an office around 9:30 and showed me a cubicle with a desk. I had no chair, no phone, no computer. I looked around the cubicle and asked, 'How will I do any work?' He shrugged and then told me that was my problem to figure out, adding that my manager was busy in a meeting.

"So, I borrowed a chair from the guy next door, left a note on my manager's door, and read the newspaper until it was time for lunch. My manager finally came by at noon and invited me out to lunch. By the time we returned, I had a phone, but the chair was gone. I turned to my manager and said, 'Hey, I'm here to work. Tell me how to do that without any equipment!'

"By 2:30, I had a phone and a computer, but still no chair, so I knelt on the floor to use the computer. It didn't take me long to realize they hadn't activated my password for access. And, since I didn't have e-mail access, I couldn't look up the phone number for the system administrator, to ask for access. I just gave up and went home.

"I came in at 9:00 the next morning, and sat in my manager's office until he was done with a meeting. I asked him for the names of people who could help me get a chair and network access. I stalked them around the company all morning and through lunch. By 2, I finally had a chair and network access. Everyone I approached was lovely and apologetic, but I just wasted two days before I even started to work. What the heck kind of place is this, anyway?"

—New hire, describing her first days on the job

Congratulations. You hired a candidate, and she starts on Monday. Now is the time to plan ahead. Ask yourself, What will she think at the end of her first day? Will she have the kind of initial experience that will make her feel that our company indeed is the very best place to work, or will she be disappointed with the company and organization?

How your new hires will answer those questions depends on how well you prepare for their all-important first day. Before I elaborate, let me draw an analogy to which most everyone can relate: The initial experience a new hire has during the first hours working in a new organization is a bit like the experience many people have when they begin installing a piece of software they have just purchased—for some software, you click a few keys, and the software installs itself and works. For other software, the installation stumbles, requiring you to load yet more software, upgrade an application you had no idea was necessary, or worse, the installation keels over and dies. My point is this: The first thing a new hire sees after your hiring pitch has ended is how good or inept your organization is at "installing" him or her. Make the first day a good one.[1]

The responsibility goes beyond the first day, however. The harder it is for your new employee to succeed during the first days and weeks on the job, the harder it may be for your company to retain him or her. An early negative impression is especially difficult to overcome.

Let's look at the investment made to date. You've spent at least one person-week—and most likely closer to four person-weeks—trying to hire this employee. If you also spent time on recruiting activities, you may have devoted close to eight person-weeks to hiring. That's a dollar investment of somewhere between $8,000 and $16,000. If you paid a contract recruiter, you've forked over an additional sum equal to one-third or more of the employee's annual salary. Clearly, your investment is sizable even before the new hire shows up for Day One; don't throw that money away right at the start by making your new hires regret their decision to work for you.

Prepare for a smooth transition *before* the new hire starts.

The preceding should have convinced you (if convincing still was needed) that your hiring work is not over just because a candidate has accepted your offer. In fact, your work isn't complete until your new hire has actually started work, is settled into your company's culture, and understands

---

[1] For a variety of excellent tips on new-hire orientation, see Joseph Rosse and Robert Levin, *High-Impact Hiring: A Comprehensive Guide to Performance-Based Hiring* (San Francisco: Jossey-Bass, 1997), pp. 282ff.; Del J. Still, *High Impact Hiring: How to Interview and Select Outstanding Employees* (Dana Point, Calif.: Management Development Systems, 1997), pp. 217ff.; and Carol A. Hacker, *The Costs of Bad Hiring Decisions & How to Avoid Them*, 2nd ed. (Boca Raton, Fla.: CRC Press, 1999), pp. 157ff.

how to perform a basic set of tasks. You—or someone who works with you—have activities to see to before the new hire starts work.

When a new hire starts, he or she will need to be able to enter your company building and office space. At a minimum, you'll need to order any necessary badges, keys, or key cards, and choose a place for the new hire to work. You'll also want to stock the newcomer's work space with a desk and a chair, a wastebasket, a lamp or two, a working phone, a computer, and miscellaneous items such as stationery, scissors, a tape dispenser, pens, pencils, and pads. In addition, if a new hire will need to use a key, security code, or some other means to access other parts of the building or office complex, be sure to issue the necessary materials on the first day. If a person also will need specific instructions to do the job or directions to locate coworkers or administrative staff, type up a sheet of information that he or she can keep close at hand during orientation.

If you've hired more than one new person to start on the same day, make sure they *all* have places to work. Otherwise, they'll start talking among themselves and wondering why your company isn't organized enough to have chairs for all of them to sit in. Note that in the previous sentence I intentionally called the company "your company," not "their company." One of your biggest challenges on Day One is to convert the new hire's thinking from "I'm coming to work at your company" to "I'm working at *our* company."

Because your new employees were candidates up until today, and probably have been preoccupied with networking to find a job, they'll keep searching for another job if you don't help them to become integrated into the company. Day One gives you the opportunity to show them they are valued; do all you can to make their first day smooth and to get them involved in their new work as rapidly as possible.

Now that you are ready for the new hire to arrive, walk yourself through what happens when your new hire appears for work.

Identify the when, where, who, and what for Day One.

When and where. What time should the new hire come to work on the first day? Where should he or she report first? Well, if your organization offers orientation sessions for new hires, the Personnel or HR Department probably has a standard time and location for that orientation, but you'll need to communicate the information to the candidate. If you or members of your staff will provide the orientation, tell that information to the candidate prior to Day One. It is a good practice to include the start date, start time, and location in the offer letter, but if that was not feasible, don't let the information drop through the cracks. Be sure the candidate knows when and where to report for work. One suggestion to make the first day easier on yourself: If your normal work hours are, say, 8:30 A.M. to 5:30 P.M. with an hour for lunch, give that information to the new hire in the offer letter (and indicate whether he or she will have a set or flexible lunch period), but have the new hire start at 9 or 9:30 on the first morning rather than at the normal starting time. That way, if not everything is ready and waiting for your new employee, you have time to complete setting up.

Who. Determine in advance whether you'll want the new hire to ask to see you upon arrival, or will the new hire meet with someone else first? Communicate the details to the new hire before Day One. In organizations that don't have a formal orientation session, HR or Personnel Department staff members may require new employees to fill out forms before reporting for the technical work—if so, give the new hire the particulars of time and location. Sometimes, a top executive or a member of the new hire's work group will make it a point to greet a new hire when he or she starts. Tell new hires who will meet them the first day, what will happen next, and so on.

What. What supplies and physical materials will your new hire need to perform the job? Make a list of everything you think the new hire will need and then take care to make those supplies and materials available. In addition, if your new hire will need access to an active phone line, voice-mail, and e-mail, get these conveniences set up in advance of Day One. Set up generic passwords on the new hire's computer and anywhere else access will be needed, and provide instructions so he or she can change the generic password to a secure one, in order to begin work as soon as possible. Even if you have a wonderful logistics group to take care of these activations in advance of the new hire's arrival, take the time to test everything yourself. Having everything in working order both is in your best interest and, in my view, is your responsibility.

Make sure the computer is hooked up to the network, and that the new hire has a working e-mail address. If the new hire can choose his or her own e-mail address, ask that it be done early in the day, so that it is working by lunch. Most people will understand a delay of a couple of hours to make sure that their e-mail address is working properly, but a delay of an entire day is too long.

Prepare the new hire's work area for Day One.

You now have a new hire who's made it through whatever preparation is necessary, and who has the basic working necessities. The next step is to provide the new hire with everything he or she needs to do the work.

As mentioned above, it is a good idea to stock a new hire's furnished office with such basic office supplies as pens and pencils, paper, scissors, a stapler, staples, and staple remover, but if you don't want to do this task— or don't have time—give your new hire directions to the supply room, and let him or her choose needed supplies.

Supply the new hire with telephone and e-mail directories and with floor, building, and campus maps, as applicable. If such information is only available electronically, make sure the new hire can access them immediately upon being shown to his or her work space or print out copies for the new hire to keep. Be sure that he or she knows how to contact other people in the organization, and that all bathrooms, conference rooms, and meeting areas, and the cafeteria, if there is one, are plainly marked on the maps and floor plans you provide as handouts. Include an organizational chart along with the phone and e-mail list so the new hire knows what each person's job is. Employees who don't know who the benefits rep and the security gurus are can't ask them the questions they need answered.

Prepare another handout that lists the applications you use (and how to find them if they are not on the employee's computer). For your technical hires, this list should include such information as the names and locations of specific applications, and of process documentation and templates. It also should include instructions on how to compile and build the product, how to access the test harness, and where to find any documentation that must be read. If you have this information all on an intranet, create a local set of bookmarks for the employee's browser.

If you decide it's worth your time to develop an information packet, either to include with your offer letter or to provide with a welcome letter when the new hire reports for Day One, you can put information such as the following in the packet:

- maps and physical plant details, showing where to access the building, building hours, parking locations, and security posts
- communications details, such as how to set up voice-mail, e-mail, and telephone access
- public and private transportation details, such as bus and rail lines, carpool points, and car services
- administrative and facilities staff contact details
- organizational charts
- tools and applications instructions
- places of interest, such as where to eat, shop, and exercise

By remembering what it's like to come in as a new employee in your company, you can put together the information you wish you had received on your first day. This might help keep your new employees in the honeymoon phase a little longer—maybe even long enough to refer other people to your company.

Explain enough of the work to help the new hire assimilate.

You've identified the basics of how the employee will work at your organization. Now it's time to orient the employee to the particular project for which he or she has been hired.

If you are not the new hire's project manager, introduce him or her to the assigned project manager. Have the project manager describe the new hire's responsibilities and deliverables, including from whom the new hire will receive direction for technical tasks (and when), to whom he or she will deliver status reports (and when), and what the status reports should contain. Introduce any other team leads and staff members to the new hire.

The new hire's project manager or team lead should establish a time during Day One to present a brief product demonstration that will provide an architectural overview of the product. Relevant marketing materials and documents reflecting the customers' perspective of the product can be handed out at the demo. If the new hire does not have experience with the technology, you, the project manager, or the team lead should put together a training guide to bring him or her up to speed as quickly as possible, but keep in mind that not everything can be learned on the first day. For a new hire without domain expertise, you may want to enroll him or her in courses your organization gives about your product.

If you are not the new hire's *project* manager but you are the supervising manager, explain to the new hire how you and the project manager will work together. Specify which of you will provide direction and when, what kind of information you each will supply, and how you each will evaluate the new hire's performance.

Filled with all of the preceding, Day One is a demanding day, so the lunch break can serve a double purpose: First, it allows people to catch their breath, make a phone call, and get some food; second, it provides an opportunity for people to socialize. Invite your new hire to have lunch— either with the entire project team and yourself or with a subset of the group. If the team is too big to include everyone at lunch, then organize a lunch with just the project-area staff members with whom the new hire will work. If you think he or she will be overwhelmed by even that reduced number of people, arrange to have lunch with your new hire and one or two of the people with whom he or she will work most closely. Take a few minutes before going to lunch to review the phone list with your new hire, annotating it to explain who everyone is, but be sure to give explicit permission allowing your new hire to forget who's who in the beginning. Whether you have a small, informal get-together or hold an all-project luncheon meeting, introduce your new hire to each staff member individually, so that everyone gets to know one another.

Assign a buddy.

Most new hires can use a little extra attention and help at the beginning, no matter what job they have taken on. By assigning the new hire a "buddy" to provide a helping hand, you can address his or her needs in a way that is less formal than by assigning a mentor or a coach.[2] The informality of the buddy assignment makes it more likely that the buddy actually will be called upon for help. So, what does a buddy do? A buddy can answer questions about how things work in your department or in your organization. A buddy can meet the new hire upon his or her arrival on Day One, or, if HR requires attendance first at an orientation session, the buddy can meet the new hire when the meeting with HR is over. The buddy is not just there to serve as a resource for the new hire on Day One only, however; the buddy can (and should) be available to help the new hire for a month or more. Because the commitment is for the longer term, a new hire who encounters any problems having to do either with orientation issues or with getting up to speed in the new job can ask the buddy for help.

When you assign a buddy, make sure to pick someone who wants to perform this role. A buddy doesn't have to be gregarious, but he or she should be genuinely interested in helping a new employee for the first month or so. A buddy should be technically capable, and should know how to use the applications, the configuration management system, the problem-tracking system, and such other systems and tools as will be needed by the new hire to perform the job. A buddy also should know enough about the new hire's area of the product to be able to help when asked, including whom to talk to for more information, and how to get along in your culture. A buddy who's a peer to the new hire is more likely to be asked questions than one who is not.

Technical leads, project managers, and managers of large groups of people generally should not be named as buddies, because they are too busy to be interruptible. Technical staff members have more flexibility and are likely to be truly able to help.

There are two important actions a buddy can take to help a new hire on Day One:

---

[2] See Hacker, op. cit., pp. 157-58, if the more formal idea of a coach or mentor appeals to you.

1. A buddy can identify the technical "deep water" in the new hire's *project*, showing him or her where project information is located and how to access it; the buddy also can identify the typical problem areas in the *product*. Although you may have given the new hire the bookmarks in a browser, a buddy who takes a few minutes to walk the new hire through the information will help him or her to become productive more quickly.

2. A buddy can create a list of questions and answers to address non-technical, survival-type questions new hires most frequently ask, such as, "Where is the cafeteria?" "How do I get supplies?" "Where are the bathrooms?" "Is there an assigned or discounted parking area?" "Are there employee commuter passes?" "Is there a health club?" "Is there a kitchen facility equipped with a microwave, a refrigerator, or a coffee maker?" A buddy can help a new hire get settled more quickly by anticipating commonly asked questions about whatever might be unique to the environment.

Questions do come up long after the first day, and a new hire who feels comfortable with a buddy won't feel embarrassed about asking questions. Plan to have a buddy available to each new hire for a month, at the minimum.

By the end of lunch on Day One, your new employee may be wondering if he or she will be able to remember everything once the real work begins, and may even be wondering when the real work will begin. If you have covered all the material you wanted to cover in the morning, you probably can take the new employee to his or her work space after lunch, where he or she can start learning the nitty-gritty of the project, either from training materials you've provided or as an apprentice to appointed members of the team.

Create and use a checklist for new hires.

If you hire people infrequently, you may find checklists helpful when trying to remember all the details you'll need to provide during new-hire orientation.[3] Similarly, if you hire a great many people, checklists can also help you keep track of elements to be covered on each new hire's Day One. The new-hire checklist that I use appears on the following pages.

---

[3] A simple but nicely organized Orientation Checklist is provided in Hacker, op. cit., pp. 159-60. Another good source for such checklists is Charles M. Cadwell, *New Employee Orientation: A Practical Guide for Supervisors* (Menlo Park, Calif.: Crisp Publications, 1988).

Orientation Checklist

Activities to complete once the offer has been accepted	Check off when task completes
Order a badge, keys, and key cards, as needed.	
Identify suitable office space, and verify that the space is clean and ready for a new occupant.	
Verify that the chosen office is equipped with a desk, lamp, chair, phone, and all necessary computer equipment, and that all are in working order.	
Order any needed furniture, office supplies, or computer equipment missing from the chosen office.	
Requisition an e-mail address, a voice-mail connection, and a physical mailbox.	
Activities to complete in preparation for Day One	
Stock the office with basic office supplies, such as pens, paper, pencils, wastebasket, scissors, stapler, staples, and staple remover.	
Verify e-mail access and computer hookup to the network.	
Verify that phone and e-mail directories and location maps are available; add the employee's voice-mail extension to the phone list.	
Identify the locations for all applications and templates for the employee's work.	
Supply the employee with printed manuals, as applicable to his or her work, or provide instructions for electronic access.	
Assign a buddy who can be available for the first month or so to answer the new hire's technical questions about how the team works and non-technical questions about staff, neighborhood, rules, and traditions peculiar to the specific environment and culture.	
Prepare a welcome letter and orientation package, including all HR forms.	
Activities to complete when the new hire arrives on Day One	
Add the employee's name and title to the organization chart; add the name and extension number to the phone directory and other relevant lists.	
Introduce the new employee to project members, executives, personnel, administrative staff, as needed.	
Show the employee instructions for calling meetings, for booking a conference room, and for other administrative procedures, as needed.	
Paperwork to collect for the new hire to fill out on Day One, to be packaged with a welcome letter and orientation packet.	
IRS, INS, and immigration forms, as applicable	
Health, dental, and life insurance forms, as applicable	
Benefits forms (long-term disability, short-term disability, and pension or retirement plans) as applicable	
A nondisclosure agreement, if applicable	
An emergency contact form	
Direct-deposit and check-cashing forms	
Business cards, as applicable	

(continued on next page)

(continued from previous page)

Paperwork to give to the new hire to keep	
Maps, floor plans, and directions	
Parking, public transportation, and commutation information	
Personnel and HR policies (conflict of interest policies; sickness, holiday, and vacation policies; lateness and absence policies; sexual harassment policies; conflict of interest policies; medical and personal leave policies; birthday lists) as applicable	

Table 13-1: Orientation Checklist

Now that you've analyzed jobs, written job descriptions, recruited staff, shepherded candidates through the interviewing process, checked references, made offers, and successfully brought new technical people on board, you may think that your job as hiring manager quite probably is done.

But wait! You still must hire a *technical manager* and you suspect that the approach you've been using to hire technical staff members won't be as effective for hiring their manager. To make matters worse, you learn that your deadlines are shorter than you thought. Don't despair. The next two chapters address just these challenges.

POINTS TO REMEMBER

- Consider what information your new hire needs to know before he or she comes to work, and provide that information in advance of Day One.
- Make a checklist for yourself as a reminder of what you have to do and when.
- Have the new hire's work space set up, fully stocked, and fully operational prior to his or her arrival for the first day.
- Increase your chances of retaining your new hires as long-term employees by being prepared for them on their first day.
- Keep enthusiastic, engaged employees happier throughout their employment by starting them off right.

14

Hiring Technical Managers

All of the ideas and steps described in this book can be readily applied to hiring technical staff members, but applying them may seem less straightforward when the candidate to be hired is a technical manager. If you've been having trouble adapting these ideas to your open management positions, this chapter should prove especially useful.

There are numerous reasons why hiring technical managers requires an approach that is different from that used for hiring technical staff members. One reason is that managers perform work that may be vastly different from the work performed by technical staff members. A second reason is that managers tend to have a larger sphere of influence than technical staff members have, and they need a wider range of experience—although they do not necessarily need expertise to the degree needed by technical staff members. Because managers perform leveraging work—work that allows other people to succeed—rather than perform the actual work itself, and because they must interact with people across a wide spectrum of technical groups, with their peers, and with the rest of the management team, it's critically important to define the management-level job as precisely as possible, so you can recruit and hire a candidate who will be likely to succeed.

In my experience, the cost of making a bad hiring decision is significant—equivalent to at least six months of the technical person's salary. Other sources report the cost can be as low as a few months' salary,[1] while still other sources report the cost at up to two to three times the technical hire's annual salary.[2] Since managers amplify the work of their staff, the

---

[1] These U.S. Department of Labor figures are cited in Carol A. Hacker, *The Costs of Bad Hiring Decisions & How to Avoid Them*, 2nd ed. (Boca Raton, Fla.: CRC Press, 1999), p. 2.

[2] For a thorough discussion of the cost of a bad hiring decision, see Lou Adler, *Hire with Your Head: Using Power Hiring to Build Great Companies*, 2nd ed. (Hoboken, N.J.: John Wiley & Sons, 2002), pp. 290ff.

cost of a bad hiring decision for managers is much higher—up to five to ten times their yearly salary.[3] Whatever the cost of hiring a technical manager who sooner or later must be fired because the job he or she does is not satisfactory, the cost is higher than the cost of hiring and then firing a technical staff member. The economics work out this way simply because a technical manager's salary almost always is higher than a technical staff member's. To avoid wasting time and money hiring an unsuitable manager, define the manager's job *fully and precisely in advance of the hiring process,* using the following steps:

1. Define the value you want the technical manager to contribute (that is, identify the problems the manager must address and solve, and describe the contribution he or she must make to the department and company).
2. Identify with whom the manager will need to interact.
3. Define the management level.
4. Define required qualities, preferences, and skills.
5. Define the required level of technical expertise and the skills required.
6. Define what activities must be performed and what deliverables will be expected.
7. Define the specific elimination factors for this management position—those factors that would eliminate a candidate from consideration.

Once you've analyzed the job to as precise a level as is feasible, write the job description to be as limiting as you can, and then continue with your normal recruiting activities.

Define the value you want the technical manager to contribute.

All managers need to address and solve general problems of management. Most managers spend time recruiting, retaining, and managing people in their organization. That kind of information about a manager's responsibilities is not what I'm encouraging you to specify in your precise and limiting job description—to the contrary, when you analyze the job you need to fill, consider which specific problems this particular technical manager will help you solve. Some examples of what specific managers might do follow:

---

[3] Adler, loc. cit.

- An *MIS manager* solves the problem of acquiring and maintaining the company's infrastructure.
- A *development manager* solves the problem of organizing and planning system or software development.
- A *customer-support manager* solves the problem of managing customer response time and customer satisfaction with product support.

All managers solve generic management problems by handling staff needs such as reviews, promotion, and the like, but you will want to determine *prior to beginning the recruitment and interviewing process* what other, specific value the manager will provide to the company.

One way to think about management value is to consider what kind of group the manager will lead. *Functional* managers, for example, organize the work of similar people. They hand off their deliverables to another group. *Project* managers coordinate the work of numerous people to deliver a product to the organization. *Matrix* managers manage people of a similar function and deliver people to the projects.

Depending on the group's structure, the manager will solve different kinds of problems. In the following table, I give an example of how different types of managers would solve the indicated problem in each organization:

Organizational Structure	Nonspecialized Manager's Area of Responsibility	Area-Specific Interactions
Functional Organization	Focus on the functional group and the value the group provides to the business	A development manager focuses on understanding the requirements and generating product to meet those requirements; also may notice process improvement needs and drive change in the function.
Project Organization	Focus on the project and the project's value to the business	A project manager (who may also be a development manager) will drive the project to completion. He or she may notice process improvement opportunities across projects, especially if the teams change among projects.
Matrix Organization	Focus on increasing the value of the functional group to the project and the business	A functional manager in a matrix organization assigns different people to each project, assisting those who need help, coaching, and mentoring where appropriate; also may notice process improvement needs and drive change across the projects.

Table 14-1: Generalist Management Interaction Analysis.

In a functional structure, the manager deals with his or her group's problems and how that group helps the business. In a functional organization, each manager hires staff members who have common experience and skill-sets, manages his or her own projects, and delivers product to other functional organizations. Many smaller MIS organizations, especially ones that are staffed by people who acquire, modify, and manage software used internally, are organized functionally. The smaller the staff and the more functional the organization, the more technical expertise the manager needs to solve his or her functional group's problems. Managers with multiple function responsibility need project management expertise, as well as the business expertise to understand the issues of the project.

In a project structure, the manager deals with his or her projects in terms of how the assigned projects contribute to the business. The manager gathers people of varied functional expertise (developers, testers, and writers, for example) who work together on a project. The manager delivers the product to the organization, not to another project group. These managers need to know how technical products, in general, are created and delivered, but they don't always need to understand the specifics of how the actual product is created.

In a matrix structure, the manager deals with the functional group's issues, such as how to assign and manage the group across projects, and how the functional work and the project work contribute to the business.

In a matrix organization, the functional managers assign staff members to projects, assess staff performance, and work with technical staff members to develop their careers. The project managers have responsibility for managing the projects. Here, the functional manager needs enough technical expertise to be able to assign staff members to project work and to assess staff performance. Functional managers in a matrix organization may need project management expertise in addition to business expertise.

Once an organization reaches a size of twelve people or more, the work tends to take on the characteristics of either a project-based or a matrix organization, whether that structure is formalized or not. When defining the management problem you want to solve, look at your organization's informal structure as well as at the formal structure, to see how much breadth of management skill this manager will require. A manager in a functional structure requires more technical depth than, say, a manager in a matrix structure, whereas a manager in a matrix structure requires considerable breadth but less technical knowledge.

Knowing the problems the manager will solve helps you define the interactions the manager will have at work.

Define the technical manager's interactions.

A technical manager's interactions are more diverse than those of the technical staff. Although technical managers interact with their groups, the question you'll need to address when hiring is how much diversity you require of this manager's interactions.

Because technical managers work within their departments, across departments, and at numerous levels in the organization, they use cooperation, influence, work direction, and negotiation skills every day. But each manager will require different expertise in these interactions.

First, look at the technical staff already in the group. Many technical staff members will challenge their managers' decisions and direction, and most will expect to have a manager who will stand up to them.

Once you've reviewed who will be on the technical manager's staff, look at the other people with whom this manager will work, and then evaluate their ability to cooperate with their peers as well as at their ability to manage people working at management levels above their own. For example, some product managers push back on development managers. Some development managers challenge the test managers. Some line managers need to be able to guide their staff members so as to meet the expectations of senior management while still protecting them from senior-level (or other) management's interference and sometimes impulsive direction.

Use the structure of the organization to identify with whom and how intensively the technical manager will interact:

Organizational Structure	Technical Manager's Area of Responsibility	Area-Specific Interactions
Functional Organization	Focus on the hierarchy of the functional group	A development manager focuses on his or her own group and up toward the company's CEO or Director. Infrequently, he or she presents project status details to external stakeholders.
Project Organization	Focus on the project and up into the manager's reporting hierarchy	A development manager plans and executes the work to be done by the project, reporting progress to his or her manager and to other stakeholders, who may possibly be external.
Matrix Organization	Focus across projects as well as across the larger functional area	A development manager assigns people to projects, working with project managers and other functional managers when problems are reported. He or she may report progress to external stakeholders.

Table 14-2: Technical Management Interaction Analysis.

Once you have identified the technical manager's expected types of interactions, you can define needed skills and experience.

A True Story

One company I was asked to diagnose had had trouble retaining its test managers for longer than one year. After the company had gone through three test managers in as many years, the hiring manager invited me in to help determine why this was happening. After talking with staff members and management, I discovered that when the test managers were hired, they met exclusively with members of the test team—a considerate, empathetic, and consensus-driven group—but that when they began working at the company, they needed to interact frequently with the two development managers, who unfortunately thought the best way to deal with their differences was to argue about each conflict. These battles were waged loudly enough so that everyone on the floor knew who was taking which side.

After we talked about what I had identified as the probable cause for such turnover, the company changed the interview strategy, and began including the two fractious development managers on the interview team. The first two candidates interviewed by the new interview team walked away from their interviews and withdrew their applications for the job of test manager, explaining that they didn't wish to work in a place where they would need to yell about everything.

Rather than attempt to change the two development managers' argumentative style, the company's management realized it needed to look for people who could work with both the development managers and the amenable, agreeable testers. Once the need to work with both kinds of people was made a transparent part of the interview experience and was understood as being a criterion of employment, the company was able to hire *and retain* a test manager suited to the job and the environment.

Define the management level.

The management level will define what combination of strategic and tactical work the manager will be expected to perform. As you consider what kinds of strategic, operational, and administrative work the manager will do, keep in mind that the higher the percentage of strategic work needing to be done, the higher the manager's level will need to be.

Management level is determined by a combination of factors. To determine the level, consider who the manager will work with—and at what levels—across the functions of the organization, how much negotiation skill the manager must have, how persuasive or influential the manager must be at convincing people to do as directed, and how many people the manager will be expected to oversee.

If the manager will need to manage multiple groups at multiple levels, make sure the title you post for the job opening suits the diverse population and that it reflects a position high enough up the corporate ladder to garner the respect of those to be managed. One way to know whether you've defined a job with management responsibility at too many levels is to look at the title. If you have to use more than one phrase to define the title, you've defined the job too broadly. Over the years, I have come across some titles that I suspect attempted to encompass too many levels. If you post a job title as "Director of Quality and Support Manager," "Director of Development and Release Engineer," or "Director of Program Management Office and Development Manager," you most likely will find that you are looking for the impossible. Let's think about this: The first and third titles seek someone who combines the role and responsibilities of both director and manager. The second title neglects the fact that someone who directs a development effort works with different skills and at different levels than someone who is trained as a release engineer.

Several years back, an organization notified me that it was looking for a "QA Manager," but in truth, what it really wanted was a person who could develop tests, lead the testing effort, lead the process improvement effort, deal with a second-line support group, and jump over tall buildings in a single leap. Even the most talented superhero of QA managers wouldn't stand a chance of succeeding in a job with so many levels of work. My point: Make the job title match the work.

By examining each component the organization hoped to find when it posted the job title as "QA Manager," we can break down the work needing to be done, ranging from the most strategic to the most tactical:

Most Strategic Work	Lead process improvement effort.
⇕	Manage testing effort. Manage second-line support group. Define metrics to gather for test and support.
	Develop test plans. Develop support group plans.
	Gather metrics for test and support. Report on those metrics.
Most Tactical Work	Develop tests.

Although it is highly unlikely that one manager could work successfully at these multiple levels, we can dissect the tasks to find how best to get the work done and then we can assign the logically recombined tasks to qualified people, as follows:

- One manager might both develop tests and lead the testing effort because the effort is focused in the specific functional area of testing, and is to be performed at one tactical level, testing a particular product; or
- One highly experienced manager might lead the testing effort and the process improvement effort, and also oversee a second-line support group because all of these management activities are at a common strategic and tactical level; or
- Different managers may need to perform the work of gathering metrics (predominantly administrative) from those managers who must define what metrics are required and how to gather them (strategic).

Clearly, one solitary manager can't perform the tactical and administrative work, the testing, and the gathering of metrics in addition to the other required management activities. Because the activities are not grouped logically, the work requires the manager to be involved with the product, the process, peer managers, senior-level managers, and possibly even with customers at different levels of responsibility.

Few managers can think well at multiple levels because the types of problems they must solve are too diverse. Consequently, they choose a level and work there, whether that's the level you want the manager to work at or not.

Managers don't just have problems handling a variety of strategic and tactical levels. They also may have trouble dealing with technical responsibilities in addition to their management roles.

Many first-line managers do have technical responsibilities as well as management responsibilities. However, every manager still performing technical work must consider how much time needs to be spent on technical issues and how much time on management. Before you assign yourself the task of defining requirements for a management position to be filled by someone capable of both managing and making a serious technical contribution, consider what author and consultant Tom DeMarco calls The Second Law of Bad Management: "Put yourself in as your own utility infielder."[4]

---

[4] Citing the action a manager takes when he or she leaves the management role untended and pitches in to work shoulder to shoulder with staff, Tom DeMarco calls it one of the dumbest actions a manager can take, and devotes an entire chapter to the topic in *Slack: Getting Past Burnout, Busywork, and the Myth of Total Efficiency* (New York: Broadway Books, 2001), pp. 80–85.

I believe that a first-line manager may be able to manage one or two people and still contribute at the technical level, but once someone has responsibility for three people or more, he or she probably can't make a significant technical contribution. Such a manager will be too busy removing obstacles, planning work, fixing things that are "broken," writing performance evaluations, and so on—in other words, busy doing the job of management. A new first-line manager will need more time to succeed at these tasks, because he or she has never done them before. If you're considering promoting a manager from within or hiring people who have never managed before, don't add technical work to their plate until you know they can accomplish both kinds of work. If the new manager already has responsibility for doing technical work, make sure that work is off the project's critical path, so that he or she also can succeed in the management role.

The presence of a technical first-line manager who performs technical work alongside the team members he or she manages can create a confusing environment for the people on the team, who don't know whether they should treat the manager as their manager or as their colleague and coworker. If you're trying to create an environment in which manager and teammates can work together in pairs or small groups, it's especially important to clarify everyone's role and to indicate to the manager and the team members that it's okay for them to work without fear of making mistakes—even in the presence of the manager.

An experienced first-line manager, particularly one who doesn't want to move up in management but who prefers to keep working hands-on, may be able to do technical work and still perform his or her managerial duties. For most people, however, it's difficult to perform both managerial and technical tasks successfully.

Many of the difficulties arise because technical decision-making is significantly different in style and in context from managerial decision-making. Because of the differences, the productivity and efficiency levels of technical managers who try to perform technical work in addition to managing can suffer substantially—primarily from task-splitting and context-switching.[5] When such individuals move from technical work to management and back again, they usually will need to reacquaint themselves with the details of whatever subject is at hand. For example, when they forget how to do something that is technical in nature but they can't take time to refresh their skill-set or memory, they may introduce problems

---

[5] For more on the effects of task-splitting, see Gerald M. Weinberg, *Quality Software Management: Volume 1, Systems Thinking* (New York: Dorset House Publishing, 1992), pp. 268-86. For more on task- or context-switching, see DeMarco, op. cit., pp. 12-21.

into the product—but when they forget to do something managerial, they are likely to introduce a people-problem.

If, as a hiring manager, you are charged with hiring a middle- or upper-level technical manager or perhaps a senior-level person with considerable experience, confirm that the manager will not be responsible for performing any technical work. Middle- and senior-level managers will be expected to perform roles that spread across the organization, and they will not have time to perform the more vertical type of technical work. With responsibilities for tasks such as tactical and strategic planning, removing obstacles between groups, managing the process, negotiating between groups, and so on, middle- and senior-level technical managers cannot be effective as members of the project team.

When you analyze the job that the manager will need to perform, define the work at a single level. Then you'll be able to judge which qualities, preferences, and skills are required.[6]

Compile a list of the desirable qualities, preferences, and skills.

Because managers focus, amplify, and augment the work of other people, the qualities, preferences, and skills of the person you hire as a manager are crucial to the success of the manager and those managed. It is important to remember to evaluate these talents in the context of leveraging other people's work.[7]

It is my opinion that a technical manager should possess the following qualities if he or she is to be successful:

- *Integrity:* I want to know that a technical manager is both honorable and honest. Managers with integrity attract good people, and make it possible to extricate the people, the project, or the business from difficult situations.
- *Initiative:* I expect a technical manager to have a high degree of initiative—whether in defining the work his or her group will perform, in defining how the group should perform the work, or in determining the schedule for when the work should be performed. I also expect a technical manager to notice when things are not going well, and to either correct the problem or ask for help.

---

[6] For a discussion of which values matter and why, see Joan Magretta, with Nan Stone, *What Management Is: How It Works and Why It's Everyone's Business* (New York: The Free Press, 2002), pp. 194ff.

[7] For a list of desirable managerial talents, see Marcus Buckingham and Curt Coffman, *First, Break All the Rules: What the World's Greatest Managers Do Differently* (New York: Simon & Schuster, 1999), Appendix C.

- *Flexibility:* I expect a technical manager to know when to be a flexible problem-solver, whether the problem stems from managing people or from directing the course of the project. A manager who understands the need to be less flexible about certain things (such as process in a regulated industry) and more flexible about others (such as telecommuting or other personal work habits) can handle most every challenge likely to come his or her way.

- *Leadership:* I want a technical manager who has strong leadership skills but he or she does not have to be a full-blown, bona fide leader. It is important that I define what I mean by leading and what I mean by managing—they are not the same thing. "Leading" requires someone to have a vision, to articulate the vision, to plan the work to accomplish the vision, and then to follow through to make the vision a reality. "Managing" requires someone to plan the work and verify that the work happens. If you have a technical leader in the group, you may not need a manager who's also a leader.

Successful managers exhibit many of the following qualities:

- *They have the ability to create empathetic relationships with their staff.* Because management is about people, managers have to care about their people. That doesn't mean managers are touchy-feely (although they can be), but that they care about issues important to their staff members. The best managers create professional relationships with their staff, caring about them both as people and as employees, without falling into a parental or counseling role.

- *They have perspective.* Managers often need to be able to take the long-term view, to persevere with a decision, to explain what's happening to their staff, and so on.

- *They have a sense of humor.* Sometimes work isn't any fun at all. It's easier for everyone to complete the work when the technical manager can laugh at himself or herself or create some fun.[8]

- *They enjoy working with people.* Technical management cannot be performed without good communication between people.

---

[8] The importance of having fun at work is nicely promoted in David H. Maister, *Practice What You Preach: What Managers Must Do to Create a High Achievement Culture* (New York: The Free Press, 2001), p. 62.

A candidate who doesn't like working with people is not a good candidate for technical management.

- *They are strongly goal-oriented.* Managers set goals for themselves and for their staff members; some even set goals for every activity in their area of responsibility.
- *They understand responsibility and independence.* Managers are responsible not only for their own work but also for the work of their staff. Look for managers who understand how to be responsible without performing all the work themselves and for managers who are comfortable working independently when necessary.
- *They are team-oriented.* Managers lead teams, so look for a person who enjoys creating, nurturing, and developing teams.

Again in my experience, successful technical managers exhibit all or many of the following non-technical skills:

- *They possess excellent communication skills or understand the need to possess them.* Managers need to be able to communicate in various ways: in person, on the phone, by e-mail, and by memo. Decide whether you will need a manager who can make presentations to customers, to the board, or to other stakeholders; or someone who is a great writer or speaker; or merely someone who can learn these communication skills.
- *They know how to provide effective feedback.* Giving effective feedback is a specific communication skill. The manager who can define expected behaviors and can provide feedback to staff members in an effective way empowers the technical staff to work more efficiently.
- *They are skilled at influencing others and they possess strong negotiation skills.* Managers who can influence others to do their bidding and who have good negotiation skills will be able to make tradeoffs within the organization that benefit their work and the work of their peers and group members. I value a manager who can get needed resources from other people, departments, or companies to accomplish the work.
- *They are accomplished problem-solvers and decision-makers.* Managers need to be able to solve problems and make deci-

sions in the face of ambiguity. When you look for a technical manager, look for someone who understands how he or she makes decisions and therefore chooses correctly more often than not. Don't look for perfection in how the manager makes choices—you'll never find a perfect candidate. Instead, look at how the candidate makes decisions, so you know if he or she is aware of personal weaknesses.

- *They are comfortable with planning, organizing, and assigning work.* A manager must be able to plan staff assignments and assign work, but he or she also must be organized enough to help other people do their work if necessary.
- *They must be good at delegation.* One of the hardest skills every manager should learn is when to delegate his or her own work. Delegation of managerial tasks helps groom others to take on management jobs.
- *They should be good at observing the "current" state.* Technical managers make strategic and tactical decisions about the work every day, considering, for example, what projects to focus on, what to do when people are stuck, and when to give people feedback about their behavior so they can improve their performance. To know what requires a management decision, the manager first has to be able to see the current state, not just the desired state.
- *They must be able to work effectively under stress.* Management, by its very nature, is stressful. I look for managers who can manage their reaction to stress because I have observed them to be more successful than those who can't.
- *They must be capable of acquiring trust and confidence from the staff.* All members of a technical manager's staff need to be certain that the manager will always be in their corner.

Managers in your organization may require other qualities, preferences, and skills than the ones I look for. One way to identify other characteristics is to observe the successful managers in your own organization to see which qualities make them successful.

Don't hire managers without the requisite talent.

Every so often, you'll meet a candidate who wants to be a technical manager for what I consider to be the wrong reasons. For example, he or she may cite the need to make more money, or a desire to control the destinies of others, or a wish to impose a particular process on a project. I recommend you eliminate these candidates as quickly as possible for, in all likelihood, they will be terrible managers. If a candidate wants a job only because of the money or title, there is no incentive for the candidate to allow him- or herself to enjoy working with people or to guide the work to an appropriate state. Technical managers manage things such as project context, organizational values and culture, and so on, not people.[9] If you meet someone who wants to completely control the process, ask yourself: Would I want to work for someone who controlled everything I did? If you would not, why would you allow your technical people to work for someone like that?

Unseasoned managers may not realize what makes a manager successful and, in attempting to explain why they want to be a technical manager, may use phrases such as "I want to supervise people," or "I hope to impose discipline on the process." Those phrases are not necessarily predictors of bad managers, but they may well be. For such people, you'll have to ask them to explain what those phrases mean to them.

Define the manager's required technical expertise.

Every management job is different, so define what the manager needs to know to perform the particular job of management for the group and your organization.

If you're trying to hire a first-line manager, the manager needs to understand the process your organization uses to create your products, and enough about the technology to assess risks in the product and the process. A middle-level manager needs to understand how your organization's products are used in your industry (or by your organization), and how the group's work affects the rest of the organization. A senior-level manager needs to be able to lead strategic planning for the business, and needs to determine which products to create to support that business.

Too often, we assume that managers of technical people need to be technical themselves. But you, like me, have probably seen talented technical people turn into lousy managers and non-technical managers turn into great managers of technical people. If you're not sure how much technical background a manager should have, consider four areas of experience, as follows:

---

[9] For the point of view that it is "management's responsibility to provide a context of values within which individuals can manage themselves, and . . . what it means for individuals to take responsibility for their own performance," see Magretta, op. cit., p. 195.

- *Functional-skills experience:* Normally, technical managers don't need to be the technical gurus in their functional area; rather, they need to understand how to *hire* people who are the technical gurus in that functional area. For example, the manager may need to know that the project requires configuration management or a particular kind of testing, but not how to perform the functions in configuration management nor how to test. Consider your organization's product line and already existing expertise before you decide that the manager needs to know everything about development, testing, support, or other function—or nothing. If you're convinced a manager requires a technical degree, define which functional skills the manager will use in this management job.

- *Domain experience:* Managers need to know how customers use the product; the more they understand about how the product works, the easier it will be for them to make decisions. In addition, the manager needs to understand how his or her group contributes to the creation of the product or service (the product development, support, and testing processes, for example). The more the manager understands about the functional group's process, the better his or her steering and decision-making abilities will be. For middle- and senior-level managers, process knowledge is essential. Normally, managers make most of their decisions in the product domain, or in the functional group's process domain.

- *Tools and technology experience:* Managers need a small amount of expertise when they are making low-level decisions about technology. Managers who work as architects, choosing product architecture, need in-depth knowledge of the technology. However, most managers only need to be able to contrast different technical approaches to problems. To find a manager with necessary experience and expertise, you'll need to define the management tools the manager should know how to use, such as project-scheduling, project-portfolio, or spreadsheet software. Then, define how much of that technical experience can be learned while the manager is on the job, and how much is a prerequisite for hiring.

- *Industry experience:* The more the manager knows about the industry, the better the manager's decisions will be, especially if the manager is making strategic decisions. For managers hired by telecommunications vendors, it's relatively easy to acquire industry expertise—much of the necessary information is readily accessible. Regulated-industry expertise may be more difficult to acquire, depending on the area within the industry. For example, a software development manager who has never worked for a pharmaceutical company may not realize the necessity of requirements traceability until too late in product development.

Determine a manager's technical experience and expertise in the areas of highest risk for the problems you want solved.

Define which activities and deliverables the manager will oversee.

Most technical managers have some responsibility for both individual and shared activities and deliverables. What work do the managers in your organization need to perform? Possible tasks technical managers must be able to perform include the following:

- *They must manage the project portfolio.* Specifically, they must evaluate all the work to be done, including project work, periodic work, maintenance work, and the unique tasks, to determine project priority and staffing assignments.
- *They must assign the work.* They must be able to determine who should do which work and when.
- *They must hire and retain staff.* The job will require that they hire people, give staff members timely feedback, develop and deliver performance evaluations, distribute rewards, and give recognition for work done by others.
- *They must work with peers.* Can they successfully negotiate for resources, influence others to achieve the best results for the organization, and provide an environment that is conducive to productive collaboration?
- *They must tend to project administration.* They may need to develop capital equipment and expense budgets and reports.
- *They must assess and spearhead process improvement.* Since the manager is the first to see where the work is headed and since he or she has the ability to see the whole organizational picture, the manager is frequently the driver of process improvement.

- *They must communicate effectively with others in the organization.* It is a plus if the manager can communicate with a wide range of personalities and people (not just with his or her team or peers, but with the entire organization).
- *They must make decisions in the face of inadequate information.*

Managers perform more or less of this work, depending on their level and on the organization in which they work. Consider which activities—with their associated deliverables—the manager you'll want to hire must be able to perform.

POINTS TO REMEMBER

- Define the problems you want a manager to solve. Then you can complete the job description and analysis.
- Define the type of technical expertise required. If you find that you're looking for someone with more experience with the technology, decide whether you're looking for a technical manager or a senior-level technical person.

15

Moving Forward

"I analyzed the job, and wrote the job description. During the past three weeks alone, I screened sixty résumés and interviewed five people. I just can't find anyone who fits. I've spent three months, and now I'm two weeks away from the time I need the new hire. If I don't find someone now, I won't be able to staff the project-to-save-the-company."

—Frantic development manager

Take action to fill your open position even when no one seems just right.

First, review your job description. Are you placing too much emphasis on tools or technology skills or otherwise over-constraining the job description, and too little emphasis on a candidate's ability to learn how to use the new tools or technology once he or she has been assimilated into your environment? If this is the case, make sure you are using all available strategies for recruiting the best candidates. If you need to recruit someone with a rare combination of preferences, qualities, and skills, then make sourcing your top priority and use multiple sourcing strategies.

When you come across a job you can't seem to fill, you need to recheck the steps you've been using. Take the time first to verify that you have correctly defined the job—that is, verify the job analysis; the compensation package; your résumé-screening filters; and your interview team's procedures, questions, and approach. Once you are sure that these methods are not the reason for your current dilemma, you can proceed in any number of directions, as follows:

- You can choose to wait.
- You can hire from within.

- You can hire a candidate with some skills and train him or her until the rest of the required skills are acquired.
- You can hire a contractor rather than a permanent employee.
- You can replan the project.
- You can change the way you've assigned people to projects.

Whatever you do, don't hire the wrong candidate just to put a "warm body" in the position, or because the requisition is on the verge of disappearing. The wrong hire is always a bad hire.

If you're not under project pressure, you may be able to continue to wait for the right person to come along. In order to make a reasonable decision, however, you may need to refine your hiring strategy.

Verify that your hiring work is on track.

When you have trouble hiring, ask the following questions:

1. *Is the hiring strategy still accurate?* If you completed a hiring strategy template according to the details provided in Chapter 1, "Developing Your Hiring Strategy," now's the time to review it. Perhaps you originally thought the opening to be filled would add new skills to augment the skills your current staff members already have, but now, after reviewing some résumés, you realize your strategy should be to transform your group or possibly even to create an entirely new group.

2. *Is the job analysis still accurate?* If you haven't reviewed and modified the job analysis to expand or change it to better suit the job opening, look for a pattern as to why you've rejected each candidate. Determine whether the candidates were missing some quality, skill, or preference that you had identified as necessary when you prepared the analysis or whether you were looking for other preferences, qualities, or skills that none of the candidates had.

3. *Are you using a wide range of sourcing techniques to find candidates?* If you're only using one or two of the inexpensive sourcing techniques, you may never find suitable candidates. During economic boom times, you may need to use at least four sourcing strategies to attract someone to the job.

4. *Is the compensation package attractive?* If you're not able to pay market value to the candidate, work with your management and with Personnel or Human Resources staff members to come up with an offer a qualified candidate can accept. If salary was the primary reason candidates would not take the job, you may be able to offer attractive perquisites in addition to salary.

5. *Are you screening candidates too rigidly while reviewing résumés or during the phone-screen*? If too many suitable candidates fail your phone-screen, review your phone-screen questions with your interview team and any recruiters you're using.

6. *Are your interviewers' questions appropriate?* Review the types of questions interviewers typically ask, and inquire about any ad-libbing they may be doing. You may find that candidates have been turned off by something someone on the interviewing team has disclosed. Or you may discover that your interviewers' questions are not providing enough information for the interview team to make a good decision.

7. *Do you see patterns of rejection?* Are you inadvertently rejecting suitable candidates? If so, determine why and fix the process.

If you've reviewed all your work and it seems right, you do have other options.

Know how long you can wait for the right candidate.

You may be able to wait to hire if, even before the work is to begin, you've planned both what work will be done and how you'll staff the work. If you're the hiring manager but not the project manager, talk to the project manager to know how long you can wait to hire someone. Determine the risk of waiting to hire, and the effect waiting will have on the project manager's plans.

If the project manager hasn't planned far enough ahead to let you know whether or not you can wait to hire, work with him or her to define the critical paths and interdependencies. Once you know what the risks, critical paths, and interdependencies are, you will know how long you can wait to hire. If you wait too long, you may doom the project to failure.

Before you reach the can't-wait, project-doom date, select a new hiring strategy. One technique is to hire from within the organization.

Hire from within the organization.

When you have a large enough organization, hiring from within can be a reasonable choice. You can post the job description on your company's Website or job board, actively recruit from other departments or divisions of the company, or promote people in your own group.

Hiring from within, especially if you can promote someone, has one substantial benefit: Your staff members can see that you're interested in people's career paths. By promoting people to better positions, you are advancing their careers, not just filling open positions. In return, it is probable that your staff members will feel greater loyalty to you and to the company, and they may even be motivated to take on increased responsibility.

In addition to garnering loyalty, promoting people from within also demonstrates that you value people who already understand the company, the products, and most importantly, your organization's work culture.

If you do hire from within, you may have to replace the person who has moved up or out, but this may be easier to do than filling the open position from the sorry pool of candidates from the outside. Sometimes, you may find yourself with an opening because another manager within your company has offered your employee a better position. If that happens, encourage your employee to move on, and prepare to replace him or her. Whatever you do, don't demand that the employee remain in your group until you've found and trained a replacement. Employees are not chess pieces that you can retain or move capriciously. It should be the employee's decision whether or not to stay to help train his or her replacement. Asking an employee to work part-time in your job and part-time in the new job until you find a replacement is not a good idea either. Don't try to share employees; give them up.

If you offer an employee a new role in your group or a move to a new position within another group, you can use your consensus-based hiring process and behavior-description interviewing techniques, canvassing the candidate's current manager and peers as references. You may find that a candidate looking for a different role within the current company will be more open about his or her involvement in a job search than a candidate who is looking outside the company. You can benefit from such openness because it will allow you to invite honest discussion and dialogue with a variety of staff members and managers.

Hire a candidate with limited skills if he or she can be trained.

Sometimes, you will find someone who almost fits the job description, but isn't a perfect match in all regards. In such a case, you may want to consider hiring the candidate and then training him or her as needed.

There also are real benefits to be derived from providing ongoing education for your current employees. On-the-job training is beneficial because it enables you to teach required skills and abilities to your staff from the inside. If you put a mentoring system in place, you can apprentice junior-level people to be trained to take the place of newly promoted or departed employees. Your current employees may lack specific technical skills but they may have valuable non-technical attributes—loyalty, team skills, dependability, organizational knowledge, or a highly refined working relationship with others on the team, for example—which you will not want to lose. You can train them in the technical skills, or provide a mentor to bring them up to speed.

If no one on your staff seems capable of learning a specific new skill, your best strategy is either to hire new people who have mastered whatever is needed or to hire new people whom you can train. I've found that some skills are fairly easily transferred or learned. Table 15-1 shows what skills I look for when I must fill a particular opening. Roles are listed in the left-hand column, and skills are described in the right. Note that skills are not necessarily related to experience.

Technical people tend to be able to learn tools and technology skills fairly quickly. For example, once a developer has learned one procedural language and one object-oriented language, it's relatively easy for that developer to learn and use other languages. Similarly, project managers who already use one project-scheduling tool can easily learn another project-scheduling tool. Testers who understand how one testing tool works can easily learn how to use another testing tool.

It's harder for such people to pick up missing functional skills without specialized training. For example, someone may know how to manipulate scripts and stored procedures for a particular database, but he or she may not understand the general theory of how databases work (that is, the person does not have functional database skills). Non-technical testers who have industry expertise may not have enough of a technical background to learn how to test your product in time for your project to succeed.

Open Position	Complementary Qualities
Software Architect	Seek someone with the ability to see the whole system, who may already work with an architect or as a high-level designer; a person who is able to discuss requirements in the context of the system's components and who has established good working relationships with the rest of the project team.
Developer	Seek someone who has demonstrated the ability to learn and to work with your organization's products or programming language. If you need someone with experience with a particular OO-language, look for alternative experience with languages such as C++, Smalltalk, LISP, or Java. Look for the way a candidate has used language at the system level close to the hardware, at mid-level between layers of several applications, or at the purely application-oriented, top level. Or, seek someone who has domain expertise in the solution-space, such as a technical support staff member or a tester.
Tester	Seek someone who is questioning, critical, curious, clever, cautious, and observant; who can take disparate events and see how they link to cause problems. Seek someone with product-domain experience, which is helpful for recognizing good tests; with software skills that are useful for understanding the code and how to test from the code; and with industry experience that is useful for knowing how customers expect the product to work in general.
Release Engineer	Seek an organized person who understands how the product needs to fit together, and who works well with others.
Project Manager	Seek someone who has served as a technical lead, organizing functional-area-specific pieces, and who can articulate goals; a technical team leader who can establish good working relationships with the project team, the functional manager, and the program manager, and who is an excellent problem-solver.
People Manager	Seek someone who has been or is a project or program manager with experience using persuasion and influence instead of authority; who can motivate, energize, and lead a group of people; and who thrives on ambiguity and is willing to make decisions regardless of insufficient data.

Table 15-1: Candidate Training Chart.

When you consider training someone to use a specific set of skills, analyze the person's background and expertise. What experience does he or she have in the areas of functional skills, product domain, tools and technology, and industry? What does the candidate not have that you need? Once you've determined how the candidate scores in these areas, you can decide whether he or she is suitable to train.

When you're willing to train, you have some advantages that you would not otherwise have. For example, you can proceed with the project because you can staff it with the right number of people even if they do not have all the needed skills at the onset. An additional advantage is that technical employees generally like to receive training, and so you will have people who are enthusiastic about their new work.

Not everything will be advantageous, however, as there are certain risks associated with hiring and training people to work on a project already in progress. For example, you probably can't maintain your schedule for deliverables and the like because time taken for training will eat into the time needed for actual project work. Another risk is that you may de-motivate your current staff members if you only provide training for a few of them or only for your new hires. An additional risk in providing training only for new hires is that you may lose valued staff members because they will not want to work for a company that doesn't give current employees training.

Hire a contractor rather than a permanent employee.

If you've been looking for a permanent employee and you absolutely, positively cannot find the right person, consider hiring a contractor who can perform the job on a temporary basis. Look for someone who can step in and perform the necessary technical work, someone who fits into the culture well enough to be successful.

You can hire a contractor for almost any technical or management position. There are contractors who can be hired to serve as tier-1 support reps all the way up through CEOs. If you intend to keep looking for a permanent employee while using the contractor, however, use the contract period to review your list of essential and desirable qualities, preferences, and skills to make sure you have an up-to-date job description.

Replan the project to fit the current staff.

The job description you've developed is based on your original assumptions of how people work in your organization, but there are other ways to staff the project, as described below:

- *Rearrange the work on the project so that different people perform different work.* Think about hiring a technical writer to reduce the time a developer spends writing user manuals, or hire a project coordinator to handle administrative tasks so as to free up your technical leads' time.
- *Use an architect or designer as a technical lead instead of hiring more developers.* If the system wasn't designed correctly originally, it's difficult for each developer to understand how to make new features or fixes fit. The more difficult it is for each developer to perform his or her work, the longer the work takes and the more people you need. An architect or a designer who can see the patterns in your system may help you leverage your current developers.

- *Use more tool-builders.* Sometimes, developers build test infra-structure to the detriment of their product development work. Sometimes, testers build test infrastructure to the detriment of their test development work. If you need infra-structure, hire tool-builders to build it. If your project team is spending too much time on black-box, GUI-based testing because no one has time to build an automation tool, con-sider hiring people who prefer to build infrastructure.

- *Organize your support staff into levels.* With levels, the people who are the most knowledgeable work on the hardest prob-lems. The people who are the least knowledgeable (or the least experienced) work on the easiest problems. It's that straightforward.

If you consider changing people's roles as you replan the project, you may be able to rearrange who does what work so that you can avoid hiring people. Shuffling tasks also can free people to do more valuable tasks.

If replanning the project is not an alternative, try rearranging the project portfolio. If you can't staff a project so that you can work the way you want to, change the way you work to fit the project staff you have. This strategy can range from changing the date a project is scheduled to start or ship to changing the project's lifecycle.

Rework the project's schedule.

Many of us have worked for organizations that had more projects in the pipeline than the technical organization could complete. If you don't protest, your superiors will undoubtedly tell you to work on all the proj-ects at the same time. If you have the appropriate staff, that's a reasonable response. However, if you don't have the right staff or have run out of time or money to hire new staff members, you'll soon realize that you can't work on all projects simultaneously. Doing so will demoralize your staff, turning some members into scarce resources, making it possible that no one will finish anything.[1]

However, you can work with your management team to rank the proj-ects, and jointly decide which projects should be staffed when. By ranking projects and choosing what projects to do in which order, fewer projects will be executed, but they will be executed faster. An additional benefit of

---

[1] The idea that people can handle multiple tasks well is debunked in Gerald M. Weinberg, *Quality Software Management: Volume 1, Systems Thinking* (New York: Dorset House Publishing, 1992), p. 284, and in Tom DeMarco and Timothy Lister, *Peopleware: Productive Projects and Teams*, 2nd ed. (New York: Dorset House Publishing, 1999), pp. 136-37.

putting each product out faster is that a project becomes staffed with more people with domain expertise.

Consider a project portfolio schedule as is shown below:

A changed project portfolio schedule would allow staff to focus on prioritized projects, as follows:

Rework the project's lifecycle.

Some lifecycles require more people on staff at certain times than at others. For example, a project using a waterfall or modified waterfall lifecycle requires that all the developers be available during the design and implementation parts of the project, and that all the testers be available during the testing part of the project. If you can't staff your people that way, then choose a different kind of lifecycle.

Iterative lifecycles, such as spiral or agile lifecycles, allow you to start with a project team and successively add and change project participants over the course of the project, especially if the later changes do not affect the work that is already complete. Incremental lifecycles, such as staged delivery lifecycles as well as some new iterations of agile lifecycles, allow you to add more people as the project progresses, assuming that people are focused on a particular set of work and not the entire project.

Change the work practices.

Sometimes, project managers demand to have more testers added to the staff because they know that they run the risk of releasing a product with unknown defects if testing is insufficient. Or, an IS manager may require additional people to maintain released products. Or, the support manager may demand that additional support staff be added because the product is likely to release with too many defects.

Instead of adding testers, a project manager can change work practices to have a dramatic effect on the release date and on the quality of the product after release—a change that will affect how many people will be needed during other project phases. For example, a project that requires a peer review of each fix will face less rework at the end of the project, reducing the need for an army of testers. Alternatively, assigning a release engineer to manage the build process so as to create nightly builds, and to test builds with a smoke test, can free up developers.

The more you can do to reduce the overall number of unrelated tasks technical staff members perform, the more they can focus on their high-priority tasks, and the faster they can complete their best work.

Change the job description.

If you've chosen any option other than waiting for the right person, you now should reanalyze the job and update the job description. Then, you can resume recruiting.

A True Story

Dan, a VP of Engineering in a small company, wanted to hire two testers and a program manager. He wanted the program manager to perform engineering management as well as manage the schedule, and he wanted the testers to be able to talk to developers because the requirements were incomplete.

Dan had two hiring problems: First, the person who would fill the program manager position would need to be the rare combination of a person who would be able to manage the schedule while performing engineering management. Second, Dan needed to find the two testers *quickly*. He knew that the two problems were different and required different tactics: Filling a highly constrained position is not the same problem as filling a less-constrained position quickly.

As both the architect and the acting program manager, Dan was under pressure to fill the positions. In fact, he feared he would fail at everything if he didn't have the program manager on board within four weeks. Waiting for the right person was not a viable option. Hiring from within also was not viable because Dan's start-up was small and all technical staff members already reported to him.

Dan posed the hiring problem to his team members. They considered their options and then agreed to change the work practices and replan the project.

Needing to move quickly, Dan decided to appoint one of his senior developers to be a technical lead, thus easing the constraint that the program manager would need to have significant product-domain expertise. At the same time, the developers decided that they would conduct peer reviews of each other's work and that they would also pair-program, developing programs in tandem to reduce the total number of defects. That meant that the two testers Dan needed to hire wouldn't need much product-domain expertise at the beginning of the project.

Dan reworked the job descriptions and was able to find a program manager and one tester within two weeks. So that he would be able to concentrate more time on finding the second tester, Dan handed off some of his project responsibilities at the start of the third week. As a result of these changes, Dan was able to focus on finding the other tester, and successfully hired him before the end of the fourth week, greatly improving the chances that the project would be completed.

Choose your actions carefully.

If you decide to wait for the right person to come along, you face a substantial risk that your projects will be late or defect-laden, or both. Hiring an unskilled or poorly trained person may have even worse consequences. If you think you have hired suitable candidates for the staff, but in fact you have not, the project most likely will fail. What action you take is critically important.

If you choose to hire a person with only some of the necessary skills and then train that person, your action can have unintended consequences. For example, your other employees may be excited by the fact that you are willing to invest in training, and will ask for training for themselves. Think before you act.

Knowing how to solve staffing problems is part of your challenge as a hiring manager. Plan your staffing to create project success, whether it's for the Next Big Thing or for the next release of your product.

POINTS TO REMEMBER

- Make sure your job description and recruiting practices fit the position you're trying to fill.
- Review your hiring strategy to verify that you know what problems you're trying to solve.
- Don't be afraid to replan the project.
- Change the project's lifecycle to reorganize which work is performed when.
- Change who performs what kinds of testing or writing.
- Change who develops the infrastructure for a project.
- Change how the work is completed, using alternative techniques such as walkthroughs, inspections, peer reviews, and pair programming.

Appendix A

Walker Software Case Study: Hiring Multiple People

In this Appendix, I walk you through the hiring process used at Walker Software, the fictional organization we have referenced throughout the book. I begin with Walker's story, then show its worksheets and tables, then the nine steps of the hiring process.

Walker is a software-only telecommunications vendor with a staff of one hundred people. The company is beginning to come out of its start-up phase and is making money, but the management team is cautious, only hiring people who can be supported with the current revenue stream. Vijay is director of both Engineering and Operations, the product development and support groups that consist of a total of thirty people. His direct reports are Dirk, the development manager; Susan, the test manager; Ed, the tech. support manager; and Don and Tony, who are both project managers. Dirk, Susan, and Ed all are responsible for hiring their own staff.

Dirk manages eighteen people in Development. Five of the people are senior-level developers/technical leads. Eight people are mid-level developers, four are junior-level (but not entry-level) developers, and one is a release engineer. Dirk is looking for another junior- to mid-level developer.

In the test group, Susan manages four people now (one technical lead/senior-level tester, two mid-level testers who can read and write code, and one junior-level tester). She's looking for one mid-level, test automation engineer and a mid-level, black-box tester.

Ed has five people in Support: two tier-1 technical support reps, two tier-2 technical support reps, and one tier-3 technical support rep. Ed is looking for another tier-2 technical support rep.

Vijay has two project managers, but he wants to start up a third project and needs another project manager. At Walker, technical staff members are matrixed into the projects, and so the new project manager will need to be relatively senior, to deal with the issues involved in selecting and negotiating for the best-suited people for his or her project. When Walker Software's project staff members are matrixed into projects, the project managers (Don and Tony) organize and plan the daily work. The functional managers (Dirk, Susan, and Ed) confer with Vijay to determine which projects to fund. They also provide feedback to Vijay about the work their staff performs, and confirm that the staff is capable of performing the work. Table A-1 contains Vijay's analysis worksheet.

Problem Categories & Problems to Solve	Yes	No	Desired Characteristics & Problem Solutions
Staff: We need more people to do more of the same kinds of projects.	X		Technical skills, as long as enough candidates exist. If too few qualified candidates surface, focus on each one's ability to learn teamwork skills.
Technology: We're making the transition from one kind of technology, or work, or product, to another.		X	Problem-solving skills, skills learning new technology, adaptability, and cultural fit.
Technology: We're on the cutting edge of technology.		X	Adaptability, cultural fit, and ability to work in teams.
Staff: We're putting together a brand-new group.		X	Experience working, experience applying functional skills to a new product domain, and experience creating a new team and making the team successful.
Skills: We're filling in with other skills to change what we already do.	X		Cultural fit, fit with team, expertise in specific functional skills, and the ability to apply those skills to a new product domain.
Productivity: We want to make our projects finish faster.		X	Different functional skills, teamwork, cultural fit.
Staff: We need a few people now, but not forever.		X	Consider contractors with great communication skills, so we won't lose their work when they're gone.
Staff: We have to fire too many of the people we hire.		X	Verify that the interview team is composed of people who know how to interview and that they understand the requirements of the position. Use limited consensus to hire people.
Staff: Turnover is too high.		X	Review cultural-fit needs and verify that the interview questions address cultural fit.
Staff: Recruiting more people is difficult.		X	Use multiple sourcing mechanisms. Make sure the résumé-screening filter isn't too tight.
Culture: We need more diversity in our group.		X	Look for diversity in background, attitude, personality, product experience, as well as in race and gender. Look for different levels of experience.
Management: We need more management capability.	X		Look for management skills along with cultural fit.

Table A-1: Vijay's Hiring Strategy Worksheet for Engineering and Operations.

Vijay's primary problem is a simple and common problem—he needs more people to do more of the same kind of work. One person, the test automation engineer, will fill in with other skills. Another person, the project manager, will add more management capability.

Vijay knows that although he's ready for a technical writer, he doesn't have an open requisition. But, since he's doing the thinking now, he drafts the details he'll want to reference when the time comes for him to hire a writer.

Dirk, Susan, Ed, and Vijay all are looking for people with at least four years of technical experience. For each open position, they're looking for someone who's completed a variety of smaller projects. Vijay is looking for a candidate who's already managed at least two projects from start to end. The qualities, preferences, and skills are similar across group lines, but they're not the same.

Vijay tells each of his managers to try writing a job analysis for each of their open positions. When they have a first draft, they'll meet to review the analysis. Tables A-2 through A-7 illustrate what Dirk, Susan, Ed, and Vijay determine they will need. Table A-2 shows Dirk's third draft.

Template Questions	Dirk's Answers
Who interacts with this person? What roles does this person have in this job? What level is this role? What level are we willing to pay for? What's the management component?	Needs to work with developers, testers, writers, and project manager(s) as a software developer. Mid-level, no management component.
What are the job's activities and deliverables? What are the periodic deliverables, if any?	Must be able to perform high-level design (post architecture), implementation, to participate in requirements definition meetings, moderate and attend design and code reviews, contribute to smoke-test suite, and develop integration tests. Needs to manage subsystem development under the guidance of a technical lead.
Essential qualities, preferences, skills?	Needs excellent collaboration and teamwork skills. Must be adaptable and able to consider multiple designs.
Desirable qualities, preferences, skills?	Needs an excellent ability to focus.
Essential technical skills? Core technical skills?	Needs to knows C++, UNIX system calls, and UNIX shell scripts, with knowledge of data structures. Needs an understanding of the telephone industry.
Desirable technical skills?	
Minimum education or training requirements?	Needs four years' experience working as a developer, with two completed projects; a BS would also be nice, but it is not necessary. BS equivalent is okay.
Corporate cultural-fit factors?	Needs experience in the ways of a small company, should be project-oriented and desire a position of high growth and stock options.
Elimination factors?	Needs to work within our $70,000 cap for salary.

Table A-2: Dirk's Job Analysis Worksheet for the Software Developer Opening.

Susan, Walker Software's test manager, has two open positions: a test automation engineer and a test engineer. Table A-3 shows her initial job analysis worksheet for the test automation engineer position:

Template Questions	Susan's Answers
Who interacts with this person? What roles does this person have in this job? What level is this role? What level are we willing to pay for? What's the management component?	Works with developers, testers, writers, test lead, and project manager(s) as a test automation developer. Mid-level, no management component.
What are the job's activities and deliverables? What are the periodic deliverables, if any?	Contributes to test strategy for a specific product and can design and implement the tests. Attends technical design reviews, and reviews technical and marketing documentation. Writes good defect reports. Runs integration tests, develops system tests, updates system test structure if necessary.
Essential qualities, preferences, skills?	Needs experience collaborating. Must be highly curious and capable of planning automation testing approaches to a variety of test projects.
Desirable qualities, preferences, skills?	Must be able to work independently to improve the testing process.
Essential technical skills? Core technical skills?	Needs familiarity with UNIX shell scripts, must be able to read C++ code, with a working knowledge of UNIX system calls and data structures. Needs an understanding of the telephone industry.
Desirable technical skills?	Needs strong skills and the ability to moderate code reviews.
Minimum education or training requirements?	Needs four years working as a tester, developing test automation, or four years as a developer, using UNIX and scripts. A BS would be nice, but is not necessary. BS equivalent is okay.
Corporate cultural-fit factors?	Needs experience in the ways of a small company; must be project-oriented and desire a position of high growth and stock options.
Elimination factors?	Needs to accept a salary no higher than that of a mid-range test engineer.

Table A-3: Susan's Job Analysis Worksheet for the Test Automation Engineer Opening.

Now that Susan has completed the worksheet for her test automation engineer, she's ready to start the worksheet for the test engineer. There are several differences in the job analysis, because the two people will perform different work. Table A-4 shows Susan's draft job analysis worksheet for the test engineer position.

Template Questions	Susan's Answers
Who interacts with this person? What roles does this person have in this job? What level is this role? What level are we willing to pay for? What's the management component?	Must work with developers, testers, writers, test lead, and project manager(s) as a test engineer. Mid-level, no management component.
What are the job's activities and deliverables? What are the periodic deliverables, if any?	Must contribute to test strategy for a specific product; must design and implement the tests. Also must be able to review technical and marketing documentation, write good defect reports, run integration tests, develop system tests, and gather defect trends on features responsible for testing.
Essential qualities, preferences, skills?	Must have an ability to collaborate and a high degree of curiosity. Must be able to plan multiple testing approaches for the product line, and write great problem reports.
Desirable qualities, preferences, skills?	Must be capable of working independently to improve the testing process.
Essential technical skills? Core technical skills?	Needs familiarity with UNIX shell scripts and a testing knowledge of UNIX system calls, data structures. Needs an understanding of the telephone industry.
Desirable technical skills?	Needs to moderate code reviews, and read C++ code.
Minimum education or training requirements?	Needs four years' experience working as a tester, developing tests for telephony or other real-time systems, using UNIX and scripts. A BS would be nice, but is not necessary. A BS equivalent is okay.
Corporate cultural-fit factors?	Needs to be project-oriented, with a desire to work in a high-growth environment in a small company in which stock options are a possibility.
Elimination factors?	Must be able to accept a salary no higher than that of a mid-range test engineer.

Table A-4: Susan's Job Analysis Worksheet for the Test Engineer Opening.

Ed, Walker Software's Technical Support Manager, has one open position for a tier-2 technical support representative. Because Ed hasn't always reported to the development organization, he's grown accustomed to the way things are done in the Sales and Service organization. However, Vijay convinced everyone that keeping all the technical people together will help the knowledge transfer between developers and support staff. So far, Vijay seems to be right, so Ed is using the same techniques and processes for hiring as everyone else in Vijay's organization.

Table A-5 shows Ed's first draft of a job analysis worksheet for a tier-2 technical support representative.

Template Questions	Ed's Answers
Who interacts with this person? What roles does this person have in this job? What level is this role? What level are we willing to pay for? What's the management component?	Must work with developers, testers, writers, project managers, marketing managers, sales managers, and customers. Mid-level, no management component.
What are the job's activities and deliverables? What are the periodic deliverables, if any?	Needs to solve customer problems by identifying problem symptoms and problem location, and by developing tests to verify the problem and potential fixes. Needs verbal and written communication skills to communicate issues and resolution back to the customers.
Essential qualities, preferences, skills?	Must be able to focus on critical problems, to manage several customer problems concurrently, and to work independently.
Desirable qualities, preferences, skills?	Needs flexibility and adaptability to work on several areas of the product, and the ability to assess the overall support picture, helping the support manager.
Essential technical skills? Core technical skills?	Needs to be adaptable, with flexible problem-solving skills. Needs excellent phone and written communication skills. Must be proficient writing UNIX shell scripts, and C++ code. Needs a working knowledge of UNIX system calls and data structures. Must understand the telephone industry's need for up-time and a quick response to problems.
Desirable technical skills?	
Minimum education or training requirements?	Needs four years' experience working as a support engineer. A BS would be nice, but is not necessary. A BS equivalent is okay.
Corporate cultural-fit factors?	Needs to fit in a small company, be project-oriented, and able to accept a culture with high growth and stock options.
Elimination factors?	Must be satisfied with a salary no higher than that of a mid-range tier-2 support engineer. Must be available for rotating phone support.

Table A-5: Ed's Job Analysis Worksheet for the Technical Support Rep, Tier-2 Opening.

For his tier-2 opening, Ed wants someone who will be comfortable with technical users and who has the ability to dig into problems to discover the causes. He does not require significant technical skills from his tier-1 reps. The technical requirements between tier 2 and tier 3 are very similar, with the difference between tier-2 and tier-3 reps being their knowledge and understanding of the system. They do not need to have mastered the technical requirements for the job.

As we have seen above, Vijay is looking for a senior-level project manager. Table A-6 contains Vijay's project manager worksheet.

Template Questions	Vijay's Answers
Who interacts with this person? What roles does this person have in this job? What level is this role? What level are we willing to pay for? What's the management component?	Must work with developers, testers, writers, test lead, project managers, functional managers, and other managers. Mid-level, matrix management (no direct reports).
What are the job's activities and deliverables?	Must be able to perform project planning, scheduling, monitoring; to prepare monthly project status reports, weekly metrics reports; to manage concurrent projects; and to present project state to customers.
Essential qualities, preferences, skills?	Must excel at collaboration, needs to be adaptable to project changes, needs a high degree of leadership. Must be able to define a lifecycle and necessary processes for a project and follow it. Must be able to assess project risk and act on risk issues. Must be independent. Must be able to develop status presentations and present to customers.
Desirable qualities, preferences, skills?	Must be able to assist in budget creation.
Essential technical skills? Core technical skills?	Must perform project scheduling and task estimation; mentor technical staff with their task estimation. Must understand how to use the configuration-management system to manage code base, and must be competent in understanding the UNIX and C++ development environment and projects.
Desirable technical skills?	
Minimum education or training requirements?	Needs two years working as a project manager, plus four years as part of the technical staff of a project. A Bachelor's degree would be nice, but it is not necessary. A BS or BA equivalent is okay.
Corporate cultural-fit factors?	Must be experienced at fitting into a small company, and must be project-oriented, with a desire to work in a high-growth environment. Must be able to take stock options in lieu of a higher salary.
Elimination factors?	Must be able to accept a salary no higher than that of a senior project manager. Must have highly refined presentation skills. Must be available to travel once a quarter to a major customer.

Table A-6: Vijay's Job Analysis Worksheet for the Project Manager Opening.

On the next page, Table A-7 shows Vijay's first draft at defining what he wants in a technical writer, prepared in the event that he eventually may be able to hire a writer.

Template Questions	Vijay's Answers
Who interacts with this person? What roles does this person have in this job? What level is this role? What level are we willing to pay for? What's the management component?	Must work with developers, testers, writers, and project managers. Mid-level, no management component.
What are the job's activities and deliverables?	Must be able to refine requirements with technical staff and marketing, develop documentation, and participate in book design work. May need to design small or independent books. May need to participate in functional or user interface design.
Essential qualities, preferences, skills?	Must be able to elicit implicit requirements, to work with developers, and to work independently.
Desirable qualities, preferences, skills?	Needs flexibility to work on several areas of the product.
Essential technical skills? Core technical skills?	Needs excellent written communication skills. Must be able to understand system design in order to help define tutorials. Must understand how to function with high performance, high availability systems, in order to understand documentation deliverables.
Desirable technical skills?	
Minimum education or training requirements?	Needs four years' experience working as a writer. A degree beyond high school would be nice, but is not necessary. A junior college or some college credit in journalism, technical writing, or English would also be okay.
Corporate cultural-fit factors?	Must be okay with working in a small company. Must be project-oriented, capable of high growth, and accept stock options in lieu of a higher salary.
Elimination factors?	Must be able to take a salary that is no higher than that of a mid-range technical writer.

Table A-7: Vijay's Job Analysis Worksheet for the Technical Writer Wish-List.

Now that we have seen the worksheets and analyses for Walker Software's current hiring effort, it is time to go through the nine steps Walker follows for hiring.

Step 1: Walk through the job analysis.

Vijay asks each manager to walk through each job analysis, to make sure no one's missed anything. In Walker Software's small organization, each person will encounter every other person, so it's worth a few minutes to make sure no one's forgotten anything. The managers are all pleased with the job analyses, and proceed to their next task, that of writing job descriptions, shown in Step 2.

Step 2: Write the job description.

Using the analyses they presented in the preceding tables, Dirk, Susan, and Ed are ready to prepare their job descriptions. Table A-8 contains Dirk's draft. My interpretive comments appear in italics.

Software Developer, reporting to Dirk, Walker's Development Manager

Generic requirements:

- Minimum of 4 years as a developer, on at least two completed projects
- 2 years of C++
- 2 years of UNIX system calls, data structures
- BS, CS, or equivalent experience. (*Dirk doesn't care whether his candidates have a degree.*)

Specific requirements:

- Understanding of telephony industry. (*Dirk hasn't yet explained why this understanding is necessary or what part of the industry the candidate should understand.*)
- At least 2 years of work on the XYZ subsystem, with responsibility for implementation and unit-testing on completed projects. (*Dirk is looking for someone who's already learned how to design, and who now is learning how to develop the system's architecture. He wants someone who understands how to maintain and add onto code that may not have been sufficiently designed or tested. Note that Dirk hasn't asked for what he wants; he's using this explanation of experience as shorthand for what he really wants. For a first-draft job description, this is okay.*)

Primary responsibilities:

- Assist with requirements and overall architecture; design, implement, and unit-test the subsystem
- Moderate and attend design reviews
- Moderate and attend code inspections
- Implement smoke tests for the subsystem
- Develop integration tests for the subsystem
- Work with developers and testers to design, develop, and unit-test the subproject. (*This addresses the high collaboration and high teamwork parts of the essential qualities.*)
- Develop multiple designs for a specific problem. (*Again, part of the essential qualities*).

- Design for performance and reliability. *(This addresses why Dirk wants someone with telephony experience, who will already understand the implicit requirements. This extra responsibility wasn't specified, but was derived from the telephony industry experience.)*

Additional responsibilities:

- Able to focus on own tasks, even when the rest of the group is working on other tasks. *(Dirk has had problems in the past with developers who couldn't stay focused on their own work, but wanted to solve other people's problems. He's not sure this is the correct way to ask for a person with "high focus" who will mind his or her own business and keep working, but he figures it couldn't hurt.)*

Other factors: None. *(Dirk does not publish salary as an elimination factor. He will find out about salary during the phone-screen.)*

Table A-8: Dirk's Job Description for the Software Developer Opening.

Dirk isn't thrilled with how he's described his need for someone with "high focus," but he's written it, and he's clarified why telephony experience is important to him. He can refine the high-focus part later.

Dirk's job description is still vague about requiring that the candidates have an "understanding of telephony industry." In reality, any candidate's experience in which performance and reliability are critical to product success is appropriate, but Dirk isn't thinking about other industries. That's okay, he'll consider how to ask for what he wants when he writes the phone-screen and develops interview questions.

Susan's job descriptions for the two open test positions follow in Tables A-9 and A-10.

Test Automation Engineer, reporting to Susan, Walker's Test Manager

Generic requirements: *(These are the minimum education or training requirements applicable to this job level at this company and are not industry-specific.)*

- 4+ years as a tester developing test automation, *or* 4+ years as a developer using UNIX scripts
- Understanding of C++
- Understanding of UNIX system calls, data structures
- BS, CS, or equivalent experience. *(This is where Susan will examine the candidate's understanding of data structures and product architecture. Even though Susan doesn't care whether her candidates have a degree, her Human Resources rep cares, so Susan uses "equivalent experience" to avoid screening out people who might be appropriate but don't have a degree. Susan's HR rep is using the degree requirement as shorthand, to determine whether a candidate is qualified.)*

Specific requirements: *(Susan wants someone with strong people skills along with the essential technical skills.)*

- Understand telephony industry. *(Susan is using telephony industry experience as shorthand. She hasn't yet explained why this is necessary.)*
- Understand the difference between white-box and black-box tests. *(While Susan was writing the job description, she realized that the automation tester would need to understand the difference between white-box and black-box tests. She'd forgotten this in her job analysis, but remembered it here. Since Susan is keeping her job analysis worksheet up-to-date, she'll add the requirement to it.)*

Primary responsibilities:

- Contribute to test strategy for a specific product; design and implement the test automation
- Attend technical design reviews, and review technical and marketing documentation
- Write defect reports
- Run integration tests and analyze results
- Develop system tests
- Update system test structure as necessary
- Work with developers and testers to develop appropriate automation tests. *(This addresses the high collaboration part.)*
- Investigate different product areas for potential test automation possibilities. *(This addresses the high curiosity part.)*

- Plan automation testing approaches to a variety of projects
- Develop tests for performance and reliability during integration and system test. *(This addresses why Susan wants someone with telephony experience, so the person already understands the kinds of tests she wants. Notice that this extra responsibility wasn't already specified, but is derived from the telephony industry experience requirement.)*

Additional responsibilities:

- Ability to moderate code reviews

Other factors: None. *(Susan isn't using the corporate cultural-fit factors or the elimination factors as part of the job description—she'll use them with recruiters, as part of the phone-screen, and during her part of the interview.)*

Table A-9: Susan's Job Description for the Test Automation Engineer Opening.

While writing the job description, Susan remembered a few things that she'd forgotten to specify in her job analysis. That's okay. Separating the analysis from the description is the same as separating the requirements from design in a project: You develop a different perspective on the problem, and you're free to iterate on both the analysis and the description.

Test Engineer, reporting to Susan, Walker's Test Manager

Generic requirements: *(These are the minimum education or training requirements applicable to this job level at this company and are not industry-specific.)*

- Minimum of 4 years working as a tester
- Minimum of 2 years testing a high performance, high-availability product
- BS, CS, or equivalent experience. *(This is where Susan will examine the candidate's understanding of data structures and product architecture. Susan uses "equivalent experience" to avoid screening out people who might be appropriate but don't have a degree.)*

Specific requirements: *(Susan wants someone with people-interaction skills along with the essential technical skills to define the specific requirements.)*

- Understand telephony industry to understand regulatory and reporting requirements
- Understand how to test for high-performance, high-availability applications
- Know how to apply numerous test techniques to the product under test

Primary responsibilities:

- Contribute to test strategy for a specific product; design and implement the test automation
- Review technical and marketing documentation
- Write defect reports
- Run integration tests and analyze results
- Develop and run system tests
- Help design and run performance tests
- Update system test structure as necessary
- Work capably with developers and testers to develop appropriate automation tests. *(This addresses the high collaboration part.)*
- Able to investigate different product areas for potential test automation possibilities. *(This addresses the high curiosity part.)*
- Able to plan automation testing approaches to a variety of projects
- Able to use existing operating-system-based tests to develop new tests

Additional responsibilities:

- Able to moderate code reviews
- Able to read C++

Other factors: None. *(Susan isn't using the corporate cultural-fit factors or the elimination factors as part of the job description—she'll use them with recruiters, as part of the phone-screen, and during her part of the interview.)*

Table A-10: Susan's Job Description for the Test Engineer Opening.

Ed's job description for a tier-2 support engineer appears in Table A-11.

Tier-2 Tech. Support Engineer, reporting to Ed, the Support Manager

Generic requirements:

- Minimum of 4 years' experience working as a Support Engineer
- 2 years of using UNIX scripts
- 2 years of C++
- 2 years of UNIX system calls, data structures
- BA or BS degree or equivalent experience. (*Ed truly only cares that his candidates have the perseverance to find the problems and fix them, and believes that a degree, technical or not, is irrelevant. He put this education/experience requirement into the job description to placate his HR rep.*)

Specific requirements:

- Understand the telephony industry's need for up-time and a quick response to problems. (*In case Walker's systems go down, Ed wants people who understand both the customers' concerns and the effect that the problem will have on the customers.*)
- Possess excellent phone and written communication skills. (*Ed wants people who aren't afraid to write to or talk to Walker's customers, and who can elicit the problem from the customers, and respond well to them. Ed is looking for someone with an ability to make the customer feel good, especially if there's a problem, and the ability to document what caused the problem, the work-around, and the solution. The job description would be better if Ed said that, but this is his shorthand for what he wants.*)
- Possess excellent problem-solving skills

Primary responsibilities:

- Solve problems for customers, for any part of the system. (*Ed wants someone both good at identifying problem symptoms and at problem reproduction, and who is also skilled at testing to verify the problem and solution. Tasks may include fixing the problem, coordinating a fix with Development and Test staff members, and fixing the appropriate code line or code section.*)
- Able to select and use appropriate and timely forms of communication with customers. (*Ed wants someone to handle problem and solution descriptions, both verbally and in writing.*)
- Able to manage several customers and customer problems simultaneously. (*Ed doesn't mean that a candidate literally should be able to think about multiple problems all at the same time, but that candidates need to be able to manage their time well enough to handle multiple problems in the same day.*)

- Able to work independently. *(Independent work separates tier-1 from tier-2 engineers at Walker. Ed doesn't have the time to directly manage everyone's work every day.)*
- Able to deal with demanding customers. *(Ed wants someone with telephony experience so that he or she understands the demands of customers with significant up-time. Ed could have said this, but telephony customers have other demands Ed hasn't analyzed yet. This extra responsibility wasn't already specified, but came from his noting the telephony industry experience.)*

Additional responsibilities:

- Flexibility about product assignments. *(Ed wants people who don't need to focus on a particular piece of the product, but who are able to learn enough about the product so that they can support many pieces of the product.)*
- Ability to see the big picture when working with one customer

Other factors:

- Available for rotating phone support. *(Ed is also not publishing salary as an elimination factor. He will check on a candidate's salary requirements during the phone-screen.)*

Table A-11: Ed's Job Description for the Tier-2 Technical Support Rep Opening.

As Ed works on the job description, he understands more about what he really wants a successful candidate to be able to do. Ed hasn't described precisely what he wants, especially in the part about managing several customers and problems simultaneously, but he's approaching a description of a person who will be appropriate for the job.

Vijay's description for the project manager opening follows in Table A-12.

Project Manager, reporting to Vijay, Walker's Director of Engineering and Operations

Generic requirements:

- At least 2 years of experience managing commercial software projects, from requirements through release. *(Vijay is looking for someone with some experience in the commercial world. He's had trouble in the past with people with no commercial experience, who did not understand the need to manage the release date as well as the defects.)*
- At least 4 years of experience working on multiple projects, in any technical capacity. *(Vijay is looking for people who are comfortable discussing technical issues with customers.)*
- Understand how to effectively use the configuration-management and defect-tracking systems to organize the code for several releases and to manage the defects. *(Vijay is sure the current tactic of supporting one or two previous releases while undergoing development of one or two new releases will continue, and he wants someone who understands the problems of supporting multiple releases.)*
- Understanding of UNIX and C++ software development projects. *(Vijay isn't looking for a technical whiz, but for someone who understands the problems the project staff members can encounter.)*

Specific requirements:

- Able to assess and manage project risk
- Able to negotiate with customers, project staff, and other managers about what to do when

Primary responsibilities:

- Plan the project, and manage the day-to-day business of the project, with primary focus on meeting the release date
- Produce weekly, monthly, and quarterly project metrics and reports
- Work with other project managers and the director to produce yearly metrics
- Collaborate with project team and customers to define reasonable release dates
- Negotiate new dates with customers when requirements change
- Choose the lifecycle that makes the most sense for each project, and manage according to that lifecycle
- Define a process for the project (which techniques to use and when), and use that process

- Create and present project status to senior management and a major customer

Additional responsibilities:

- Able to work independently. *(Vijay is willing to coach a less-experienced project manager if that person can learn to work independently quickly. Vijay would prefer to pay for someone with minimal project management experience, but he recognizes that his need for a person with strong negotiating skills is unlikely to be satisfied by hiring a junior-level project manager.)*
- Able to assess the testing process and work to improve it. *(Susan has been badgering Vijay for a few months, stating that the testing process is inadequate—and Vijay thinks she has a point. However, that's not a first-level goal Vijay has for his project managers. But if he can find that testing ability in a project manager, Vijay would be happy.)*

Elimination factors:

- Need to travel once a quarter to a major customer, with a probable two nights away. *(Vijay is not mentioning salary here.)*
- Need to present project status

Other factors:

- Candidate must be comfortable in a project-oriented, small company

Table A-12: Vijay's Job Description for the Project Manager Opening.

Vijay's job description for his wish-list technical writer is short. Although the description may never be needed, Vijay wants it ready just in case he gets the hire approved. It appears in Table A-13.

Technical Writer, reporting to Engineering Manager

Generic requirements:

- Minimum of 4 years of experience working as a writer
- Bachelor's degree or equivalent experience. *(Writers don't need advanced degrees, just writing samples.)*

Specific requirements:

- Understand high-performance, high-availability systems
- Possess excellent written communication skills
- Understand and document system design

Primary responsibilities:

- Refine requirements with technical team and marketing
- Develop tutorials and other examples
- Contribute to book content and design
- Manage multiple pieces of documentation
- Work independently

Additional responsibilities:

- Flexibility and adaptability about when to write each piece of documentation

Other factors: None. *(Salary is an implicit elimination factor. The hiring manager will verify salary in the phone-screen.)*

Table A-13: Vijay's Job Description for the Technical-Writer Job Opening Wish-List.

Vijay is sure this job description isn't enough, but he thinks having a draft now will help him obtain a requisition to hire a technical writer in the future.

Step 3: Choose a mix of sourcing strategies.

Walker Software is fortunate to be hiring during a slow economy. The company can use inexpensive techniques to source good people. Table A-14 shows how Walker might pair the people responsible for organizing the work with the sourcing techniques they'll use.

Sourcing Opportunity	Responsible Staff Members
Job fair	The HR rep will set it up. Every manager will attend.
Networking at local professional meetings	Each manager—or his or her representative—will attend local meetings of SPIN, PMI, and IEEE and will bring flyers announcing the job. The HR rep will also attend the local professional meetings to network with other hiring companies.
Walker Website	Vijay is in charge of making sure the job page is updated.
Web-based boards	The HR rep will post the ads under her account on three Websites: the local newspaper-affiliated board, Monster.com, and Techies.com.

Table A-14: Sourcing Opportunities Paired with the People Responsible for Canvassing the Source.

Vijay is concerned he won't find a senior-level project manager fast enough using only the techniques in Table A-14, so he's looking for money to pay a recruiter to help him search for the best project-manager candidate.

Using these techniques, each of Vijay's managers expects to receive enough résumés to start in-person interviewing within two weeks. If the managers don't find enough people quickly, they will add such techniques as Internet advertising, advertising on Websites, and classified ads in the newspaper.

Step 4: Write generic and tailored ads.

Vijay, Dirk, Susan, Ed, and their HR rep collaborated on the start of the ad, and then each developed his or her own job-specific paragraph. They'll use this combined notice as a flyer for the networking meetings they attend. They'll also put a link to each ad on their Website, and add more about Walker's culture. For the Web résumé boards, they'll use the employer kits and add more keywords.

> JOBS AT WALKER SOFTWARE: *Looking for a small-company, project-oriented culture that expects significant revenue growth over the next year? Join Walker Software, a fiscally responsible supplier to the telephony industry, in one of these positions:*
>
> **Software Developer:** *Software developer needed for telephony product. Work with other designers to develop and enhance Walker Software's product line. We use numerous review techniques to reduce defects, including design reviews, code reviews, and smoke tests. If you're interested in building quality software quickly, and you have at least four years of experience, including UNIX and C++, send us your résumé. Refer to position SD 102.*

Test Automation Engineer: If you're ready to automate testing for a high-performance, high-reliability telephony product, we want to talk to you. You'll need to have had at least four years of experience working as a tester developing test automation, or four years as a developer developing product. You'll be helping to develop our test strategy with the rest of the test team, and then design and implement the tests. You need to be able to write good defect reports. You'll need to attend technical design reviews, and review technical and marketing documentation. You'll be running and assessing the integration tests, developing system tests, and updating the system test structure if necessary. Refer to position TA 101.

Technical Support Representative: Our customers are smart and demanding. If you enjoy discovering what a customer's problem really is, and how to fix it, and you have at least four years of support experience, we're interested in talking to you. We're looking for excellent phone and written communication skills, as well as adaptability in solving problems. Refer to position SR 103.

Project Manager: If you enjoy managing projects in a congenial atmosphere, and you know how to manage multiple releases, we're interested in you. You should have at least two years' experience managing projects from start through release, as well as an understanding of the telephony industry. Refer to position PM 104.

Send all résumés, with the position reference, to HR Rep at Walker Software, or fax them to 999-555-XXXX, attention HR Rep."

Vijay's ad for his hoped-for technical writer, which he has prepared in advance of getting a requisition, is also ready, although it cannot be posted just yet:

Technical Writer: Help us untangle our software for our smart and demanding customers. We want technical writers who also like refining requirements and participating in book content and design. If you have at least four years of writing experience, we're interested in talking to you. We're looking for writers who take pride in the work, can work independently, and are flexible. Refer to position TW 108.

Step 5: Filter résumés.

Because it's a slow economy, Vijay and his managers expect to be flooded with résumés. They decide to review résumés together each morning for fifteen minutes. That way, if someone sees a résumé that is better suited to another manager, he or she can hand it off right then and there. If the résumés continue flooding in all day, they'll also review résumés each afternoon for another fifteen minutes. The HR rep is responsible for taking care of sending out the No-pile rejection notices. If one of the managers is inundated, the HR rep may also take on the Maybe pile. The managers will call the candidates in the Yes pile themselves.

Step 6: Develop phone-screen scripts.

Vijay asked his managers to develop their phone-screen scripts right after they developed the ads. Ed hasn't hired too many people, so he's uncomfortable telling people they don't fit the requirements for his job. Susan and Dirk don't like that part of the phone-screen either, so Susan, Dirk, Ed, and Vijay decide to practice asking questions and saying no nicely to candidates who don't fit the bill.

Here are simplified versions of their basic phone-screen scripts:

Software Developer Phone-Screen Script   Yes __ Maybe __ No __
Candidate name: _____ Phone #: _____ Date: _____

1. Let's make sure we're on the same page with respect to salary. This position is a developer position, with a salary range of [low] to [high]. We don't normally bring people in higher than the mid-point. Are we on the same page?
2. How many years of C++ and UNIX experience do you have? How many years handling UNIX system calls? How many years of UNIX shell scripting?
3. Tell me about your work in the telephony industry.
4. Are you using any data structures in your current project? Tell me about your role in defining and using those structures.
5. Tell me about the team you're on now.
6. Tell me about how you design on your current project.
7. Tell me about a time when you had to change the focus of your work. When was that, and what happened?
8. What's your current salary? What's your asking salary?
9. Have you had any recent interviews? Are you expecting any offers? In what salary range?
10. What's your availability to interview? To start?
11. Why are you leaving your current position?

Table A-15: Dirk's Software Developer Phone-Screen Script.

Test Automation Engineer Phone-Screen Script   Yes __ Maybe __ No __
Candidate name: _____ Phone #: _____ Date: _____

1. Our salary range is [low] to [high]. We don't usually start people higher than the mid-point. Are we close on salary?
2. Tell me about your current project. How did you choose what to automate?
3. How did you plan the automation? What about other projects?
4. Tell me about your team. How do you work with the rest of the test team? How do you work with the developers?
5. How do you know you've automated the right things?
6. How do you know the automation is complete?
7. Are you interviewing elsewhere? Are you expecting an offer?
8. What's your availability to interview?
9. What made you decide to look for a new job?
10. What's your salary now? What are you asking for?

Table A-16: Susan's Test Automation Engineer Phone-Screen Script.

Test Engineer Phone-Screen Script   Yes __ Maybe __ No __
Candidate name: _____ Phone #: _____ Date: _____

1. Our salary range is [low] to [high]. We don't usually start people higher than the mid-point. Are we close on salary?
2. Tell me about your current project. How did you plan the testing? What about other projects?
3. Tell me about your team. How do you work with the rest of the test team? How do you work with the developers?
4. How do you know you've tested the right things?
5. How do you know the testing is complete?
6. Tell me about a recent problem report you filed that was fixed. Tell me about one that wasn't fixed. What was the difference?
7. Are you interviewing elsewhere? Are you expecting an offer?
8. What's your availability to interview?
9. What made you decide to look for a new job?
10. What's your salary now? What are you asking for?

Table A-17: Susan's Test Engineer Phone-Screen Script.

Technical Support Rep Tier-2 Phone-Screen Script Yes __ Maybe __ No __
Candidate name: _____ Phone #: _____ Date: _____

1. Our salary range is [low] to [high]. We don't usually start people higher than the mid-point. Are we close?
2. We rotate hours on the phone. Are you available to work four-hour time slots on the phones, starting at 6 A.M. until 9 P.M., manning the phones, assuming we schedule a month out?
3. Tell me about your current work.
4. What's the most challenging problem you solved in the last month? What made it challenging?
5. How many urgent problems are you working on now? How do you prioritize the problems?
6. Tell me about a recent incident that turned into a problem report that you filed that was fixed. Tell me about one that wasn't fixed. What was the difference?
7. Are you interviewing elsewhere? Are you expecting an offer?
8. What's your availability to interview?
9. What made you decide to look for a new job?
10. What's your salary now? What are you asking for?

Table A-18: Ed's Technical Support Rep Phone-Screen Script.

Project Manager Phone-Screen Script   Yes ___ Maybe ___ No ___
Candidate name: _____ Phone #: _____ Date: _____

1. Our salary range is [low] to [high]. We don't usually start people higher than the mid-point. Are we close?
2. Tell me about the environments you've worked in.
3. Tell me about your current project. How did you plan the project? Tell me about planning versus scheduling. Tell me about choosing the lifecycle. What practices did you use?
4. Tell me about a time when you wanted some resource for your project and you were unable to get it. What happened?
5. Have you ever worked on a project that wasn't successful? What happened?
6. Tell me about your team. How do you work with people on the project team?
7. How did you estimate this project?
8. How do you know when a project is done?
9. Tell me about a recent risk that you anticipated. Tell me about handling a risk that you didn't anticipate. What was the difference? What did you learn from this?
10. Are you interviewing elsewhere? Are you expecting an offer?
11. What's your availability to interview?
12. What made you decide to look for a new job?
13. What's your salary now? What are you asking for?

Table A-19: Vijay's Project Manager Phone-Screen Script.

Technical Writer Phone-Screen Script   Yes ___ Maybe ___ No ___
Candidate name: _____ Phone #: _____ Date: _____

1. Our salary range is [low] to [high]. We don't usually start people higher than the mid-point. Are we close?
2. Tell me about your current work.
3. What kind of documentation do you write? For what audiences?
4. Do you produce on-line documentation and written documentation?
5. Describe how you worked around a tool problem.
6. How many pieces of writing are you working on now? How do you give priority to each piece of the work?
7. Have you used a source control system?
8. How do you learn about the product you're documenting? How do you obtain input?
9. Tell me about a time when you were faced with insufficient information for your writing task. What did you do?
10. How do you obtain reviews for your writing?
11. Are you interviewing elsewhere? Are you expecting an offer?
12. What's your availability to interview?
13. What made you decide to look for a new job?
14. What's your salary now? What are you asking for?

Table A-20: Vijay's Wish-List Technical Writer Phone-Screen Script.

Step 7: Plan the interview.

Each hiring manager at Walker wants people from other groups to participate in hiring and evaluating candidates, and each wants to select who from his or her staff will be involved in hiring. Once their selections have been made, the managers will meet and exchange interviewer information. If a manager needs someone else to participate, he or she can negotiate that during the interviewer assignment meeting.

Each manager has created an interviewer-assignment matrix, as shown in the following tables.

Interviewers	Dirk (Development Manager)	Vijay (Director of E & O)	Susan (Test Manager)	Steve (a developer)	Sam (a developer)	Everyone
Time	8:00-8:45	8:50-9:35	9:40-10:25	10:30-11:15	11:20-12:05	12:10-12:25
Location	Dirk's office	Conf A	Conf A	Conf A	Conf A	Dirk's office
Question Areas						Meet to evaluate the candidate
Design Skills				X	audition	a
Collaboration and Teamwork Skills	X	X	X			
General Problem-solving Skills	X				X	
Decision-making Skills		X	X			
Testing Skills		X		X		
Technical Process and Methodologies	X			X		

Table A-21: Dirk's Software Developer Interviewer Matrix.

Interviewers	Susan (Test Manager)	Dirk (Development Manager)	Cindy (a tester)	Ron (a tester)	Steve (a developer)	Everyone
Time	8:00-8:45	8:50-9:35	9:40-10:25	10:30-11:15	11:20-12:05	12:10-12:25
Location	Susan's office	Dirk's office	Conf A	Conf A	Conf A	Susan's office
Question Areas						Meet to evaluate the candidate
Teamwork	X			X		
Test Planning	X			X		
UNIX Knowledge		X			audition	
Independence	X			X		
Audition: Planning, Automation Approaches			X			
Writing Skills (defect reports, other reports)		X			X	

Table A-22: Susan's Test Automation Engineer Interviewer Matrix.

Interviewers	Susan (Test Manager)	Dirk (Development Manager)	Cindy (a tester)	Ron (a tester)	Steve (a developer)	Everyone
Time	8:00-8:45	8:50-9:35	9:40-10:25	10:30-11:15	11:20-12:05	12:10-12:25
Location	Susan's office	Dirk's office	Conf A	Conf A	Conf A	Susan's office
Question Areas						Meet to evaluate the candidate
Teamwork	X			X		
Test Planning	X			X		
Testing Approaches		X			X	
Work Independently	X		X			
Audition: Planning, Testing			X			
Writing Skills (defect reports, other reports)		X			X	

Table A-23: Susan's Test Engineer Interviewer Matrix.

Interviewers	Ed (Support Manager)	Dirk (Development Manager)	Henry (a tier-2 support rep)	Cindy (a tester)	Steve (a developer)	Everyone
Time	8:00-8:45	8:50-9:35	9:40-10:25	10:30-11:15	11:20-12:05	12:10-12:25
Location	Ed's office	Dirk's office	Conf A	Conf A	Conf A	Ed's office
Question Areas						Meet to evaluate the candidate
Manage Multiple Problems Simultaneously	X			X		
Focus on Critical Problems	X			X		
Work Independently		X			X	
Phone and Written Communication Skills		X	X			
UNIX and Telephony Skills			audition		X	

Table A-24: Ed's Technical Support Rep Interviewer Matrix.

As you will see from the following grid, the project manager interview is expected to go well past the standard lunch time. Vijay will explain to the candidate that the company won't be offering lunch.

Interviewers	Vijay (Director of E & O)	Susan (Test Manager)	Dirk (Development Manager)	Ed (Support Manager)	Cindy (a tester)	Steve (a developer)	Everyone
Time	8:00-8:45	8:50-9:35	9:40-10:25	10:30-11:15	11:20-12:05	12:10-12:55	1:00-1:15
Location	Vijay's office	Susan's office	Dirk's office	Ed's office	Conf A	Conf A	Vijay's office
Question Areas	Explain about no lunch			Offer snack			Meet to evaluate the candidate
Project Planning and Estimation	X		X				
Leadership	X	X					
Collaboration and Teamwork				X	X		
Problem-solving Skills				X		X	
Audition: Project Planning and Scheduling			X				
Metrics and Communication Skills		X			Y		
Negotiation Skills				X		X	

Table A-25: Vijay's Project Manager Interviewer Matrix.

Interviewers	Vijay (Director of E & O)	Dirk (Development Manager)	Randy (a peer writer)	Don (a project manager)	Steve (a developer)	Everyone
Time	8:00-8:45	8:50-9:35	9:40-10:25	10:30-11:15	11:20-12:05	12:10-12:25
Location	Vijay's office	Dirk's office	Conf A	Conf A	Conf A	Vijay's office
Question Areas						Meet to evaluate the candidate
Adaptability and Flexibility	X			X		
Refining Requirements		X			X	
Work Independently	X		X			
Audition: Organizing Writing for a Tutorial			X			
Writing Skills		X		X	X	

Table A-26: Vijay's Wish-List Technical Writer Interviewer Matrix.

Step 8: Develop the reference-checks.

Each manager is to develop his or her own reference-checks, allowing time to ask each person whose name has been given as a reference specific questions about the individual candidates.

Appearing on the following pages are the reference-check forms the managers at Walker developed to use as scripts.

Software Developer Reference-Check Form & Script

Candidate name: _____Date: _____

Reference name: _____Date: _____

Reference position: _____

Reference phone numbers: _____

To report to: Dirk

1. Where and in what capacity did you work with the candidate?
2. How long did the candidate work for you?
3. How long have you known the candidate?
4. How would you describe your working relationship?
5. Describe the most recent project the candidate worked on with you. Did the candidate have trouble finishing the work?
6. What issues did you have with the candidate's work?
7. If I were the candidate's manager, what advice would you have for me?
8. How quickly did the candidate learn about the product or the product line?
9. How quickly did the candidate integrate with the rest of the team?
10. Why will or did the candidate leave?
11. Would you rehire or work with this candidate again?
12. What is the candidate's current or most recent salary?

Table A-27: Software Developer Reference-Check Form & Script.

Test Engineer Reference-Check Form & Script

Candidate name: _____Date: _____

Reference name: _____Date: _____

Reference position: _____

Reference phone numbers: _____

To report to: Susan

1. Where and in what capacity did you work with the candidate?
2. How long did the candidate work for you?
3. How long have you known the candidate?
4. How would you describe your working relationship?
5. Tell me about the work the candidate did, or describe the most recent project the candidate worked on with you.

(continued on next page)

(continued from previous page)

6. What issues did you have with the candidate's work?
7. If I were the candidate's manager, what advice would you have for me?
8. How quickly did the candidate learn about the product or the product line?
9. How quickly did the candidate integrate with the rest of the team?
10. Why will or did the candidate leave?
11. Would you rehire or work with this candidate again?
12. What is the candidate's current or most recent salary?

Table A-28: *Test Engineer & Test Automation Engineer Reference-Check Form & Script.*

Technical Support Rep Reference-Check Form & Script

Candidate name: _____ Date: _____

Reference name: _____ Date: _____

Reference position: _____

Reference phone numbers: _____

To report to: Ed

1. Where and in what capacity did you work with the candidate?
2. How long did the candidate work for you?
3. How long have you known the candidate?
4. How would you describe your working relationship?
5. Tell me about the work the candidate did, or describe the most recent project the candidate worked on with you.
6. What issues did you have with the candidate's work?
7. If I were the candidate's manager, what advice would you have for me?
8. How quickly did the candidate learn about the product or the product line?
9. How quickly did the candidate integrate with the rest of the support team? The development team?
10. Why will or did the candidate leave?
11. Would you rehire or work with this candidate again?
12. What is the candidate's current or most recent salary?

Table A-29: *Technical Support Rep Reference-Check Form & Script.*

Project Manager Reference-Check Form & Script

Candidate name: _____Date: _____

Reference name: _____Date: _____

Reference position: _____

Reference phone numbers: _____

To report to: Vijay

1. Where and in what capacity did you work with the candidate?
2. How long did the candidate work for you?
3. How long have you known the candidate?
4. How would you describe your working relationship?
5. Tell me about the work the candidate did, or describe the most recent project the candidate worked on with you. Did the candidate have trouble finishing the work?
6. What issues did you have with the candidate's work?
7. If I were the candidate's manager, what advice would you have for me?
8. How quickly did the candidate learn about the product or the product line?
9. How quickly did the candidate integrate with the rest of the team?
10. Why will or did the candidate leave?
11. Would you rehire or work with this candidate again?
12. What is the candidate's current or most recent salary?
13. How many people reported to this candidate?
14. How did the candidate conduct performance appraisals?
15. How did the candidate handle "difficult" people?

Table A-30: Project Manager Reference-Check Form & Script.

Technical Writer Reference-Check Form & Script

Candidate name: _____Date: _____
Reference name: _____Date: _____
Reference position: _____
Reference phone numbers: _____
To report to: Vijay

1. Where and in what capacity did you work with the candidate?
2. How long did the candidate work for you?
3. How long have you known the candidate?
4. How would you describe your working relationship?
5. Tell me about the work the candidate did. Describe the most recent project the candidate worked on with you.
6. What issues did you have with the candidate's work?
7. If I were the candidate's manager, what advice would you have for me?
8. How quickly did the candidate learn about the product or the product line?
9. How quickly did the candidate integrate with the rest of the team?
10. Why will or did the candidate leave?
11. Would you rehire or work with this candidate again?
12. What is the candidate's current or most recent salary?

Table A-31: Wish-List Technical Writer Reference-Check Form & Script.

Step 9: Extend an offer.

The final step in the process occurs when an offer is extended and accepted, and an offer letter in completed for the candidate. Each manager will modify the standard offer letter when he or she is ready to extend an offer to the candidate. (That template is included in Appendix B, along with representative templates you can adapt to your own circumstances.)

Please note that the details given in this Appendix are meant to show how different managers can work together to fill a variety of open positions. If you are looking for only one person, or if you have the luxury of being able to focus on hiring for just one position at a time, break out the information given in this Appendix for *that one position only* and disregard details not relevant to your needs. If this seems hard to do, help can be found at my Website, http://www.jrothman.com/, where you will find downloadable versions of many of the templates, worksheets, and tables for your use as you plan your hiring strategy.

Appendix B

Templates to Use When Hiring Technical People

Many of the hiring worksheets and tables that have appeared in various chapters throughout this book are gathered in this final Appendix for handy reference. By reducing them to a generic template form and by locating them together, I hope to make them as easy as possible for you to use, to modify to serve your own hiring experiences, and to refer to whenever needed. You'll see that the first example is not in template form, however, but rather that it fully replicates Vijay's hiring-strategy worksheet, originally shown in Table A-1. My reason for replicating this worksheet is to illustrate how the material found in worksheets and tables in other parts of the book has been "sanitized" to the template form shown in this Appendix. You'll also see that although Vijay recorded problem categories and problems to solve as often on the worksheet as he thought of them, the problem categories each appear only once in the sanitized version shown as a template.

Use Vijay's hiring strategy as your starting point.

Hiring Strategy for Walker Software: Engineering & Operations			
Problem Categories & Problems to Solve	Yes	No	Desired Characteristics & Problem Solutions
Staff: We need more people to do more of the same kinds of projects.			Technical skills, as long as enough candidates exist. If too few qualified candidates surface, focus on each one's ability to learn and on teamwork skills.
Technology: We're making the transition from one kind of technology, or work, or product, to another.			Problem-solving skills, skills learning new technology, adaptability, and cultural fit.
Technology: We're on the cutting edge of technology.			Adaptability, cultural fit, and ability to work in teams.
Staff: We're putting together a brand-new group.			Experience working, experience applying functional skills to a new product domain, and experience creating a new team and making the team successful.
Skills: We're filling in with other skills to change what we already do.			Cultural fit, fit with team, expertise in specific functional skills, and the ability to apply those skills to a new product domain.
Productivity: We want to make our projects finish faster.			Different functional skills, teamwork, cultural fit.
Staff: We need a few people now, but not forever.			Consider contractors with great communication skills, so you won't lose their work when they're gone.
Staff: We have to fire too many of the people we hire.			Verify that the interview team is composed of people who know how to interview and that they understand the requirements of the position. Use limited consensus to hire people.
Staff: Turnover is too high.			Review cultural-fit needs and verify that the interview questions address cultural fit.
Staff: Recruiting more people is difficult.			Use multiple sourcing mechanisms. Make sure the résumé-screening filter isn't too tight.
Culture: We need more diversity in our group.			Look for diversity in background, attitude, personality, product experience, as well as in race and gender. Look for different levels of experience.
Management: We need more management capability.			Look for management skills along with cultural fit.

Worksheet B-1: Vijay's Hiring-Strategy Worksheet.

Modify Vijay's worksheet for your own use.

Problem Categories & Problems to Solve	Yes	No	Desired Characteristics & Problem Solutions
Staff:			
Culture:			
Skills:			
Management:			
Technology:			
Productivity:			

Template B-1: [Your Name]'s Hiring Strategy Worksheet.

Collect the templates you'll need before you review any résumés.

Before you generate the job analysis template, you may need to articulate the position's qualities, preferences, and non-technical skills. Template B-1 shows the categories from Chapter 2, "Analyzing the Job," but you should substitute your qualities, preferences, and skills, as applicable. Templates B-2 through B-10 also may be modified to suit your circumstances.

Quality, Preference, or Skill	Required	Desirable	[Your name]'s Notes *(Cite any required quality, preference, or skill specified to the job.)*
Initiative *(quality)*			
Flexibility *(quality)*			
Technical Leadership *(quality)*			
Responsibility and Independence *(quality)*			
Able to Work on Multiple Projects at One Time *(preference)*			
Goal-oriented *(preference)*			
Passion for Learning *(preference)*			
Teamwork *(preference)*			
Excellent Communication Skills *(skill)*			
Able to Handle Projects of Varying Scope *(skill)*			
Influence and Negotiation Skills *(skill)*			
Problem-solving Skills *(skill)*			
Additional Factors *(qualities, preferences, and skills)*			

Template B-2: [Job Title] Qualities, Preferences, and Skills Analysis.

Template Questions	[Your name]'s Answers
Who interacts with this person? What roles does this person have in this job? What level is this role? What level are we willing to pay for? What's the management component?	
What are the job's activities and deliverables? What are the periodic deliverables, if any?	
Essential qualities, preferences, skills?	
Desirable qualities, preferences, skills?	
Essential technical skills? *(Consider the areas of technology/tool skills, functional skills, product domain skills, and industry experience.)* Core technical skills?	
Desirable technical skills *(Consider the technical skills that are not essential, but are desirable.)*	
Minimum education or training requirements?	
Corporate cultural-fit factors?	
Elimination factors?	

Template B-3: Job Analysis for [Requisition Job Title].

[Job title]:

Reporting to [manager's title]:

Generic requirements: *(Use minimum education or training requirements that are applicable to this job level at this company, but are not industry-specific.)*

Specific requirements: *(Use people interactions along with the essential technical skills and additional desirable requirements, if you have them.)*

Responsibilities: *(Use job activities and deliverables, and essential cultural qualities, preferences, and skills.)*

Elimination factors: *(Use the elimination factors from the analysis worksheet.)*

Other factors: *(Use the corporate cultural-fit factors.)*

Template B-4: Job Description.

[Company name] is looking for a [job title].

Main attractor: *(Identify what will attract the kinds of people you want to hire, phrased in terms of corporate cultural-fit factors, product, or technology.)*

Deliverables and activities:

Essential qualities, preferences, and skills:

Contact information:

Template B-5: Job Advertisement.

[Job Title] Phone-Screen Script   Yes___Maybe___No___
Candidate name:_____Date:_____Phone #:_____

1. Let's make sure we're on the same page with respect to salary. This position is a [job title] position, with a salary range of [low] to [high]. We don't normally bring people in higher than at the mid-point. Are we on the same page?
2. Tell me something about [your job's requirement] experience. How many years of [your job's requirement] experience do you have? How many years of [your job's requirement]? How many years of [your requirement]?
3. Tell me about your work in the [your field] industry.
4. Are you using any [your requirement] in your current project? Tell me about your role in defining and using those [job skill].
5. Tell me about the team you're on now.
6. Tell me about how you [job function] on your current project.
7. Tell me about a time when you had to change the focus of your work. When was it, and what happened?
8. What's your current salary? What's your asking salary?
9. Have you had any recent interviews? Are you expecting any offers? In what salary range?
10. What's your availability to interview? To start?
11. Why are you leaving your current position?

Template B-6: Phone-Screen Script.

Interviewers	[Mgr. name]	[Interviewer name]	[Interviewer name]	[Interviewer name]	[Interviewer name]	Everyone
Time						
Location						
Question Areas						Meet to evaluate the candidate

Template B-7: [Job Title] Interviewer Matrix.

[Job Title] Reference-Check Form & Script

Candidate name:_____Date:_____

Reference name:_____Date:_____

Reference position:_____

Reference phone numbers:_____

To report to: [manager]

1. Where and in what capacity did you work with the candidate?
2. How long did the candidate work for you?
3. How long have you known the candidate?
4. How would you describe your working relationship?
5. Describe the most recent project the candidate worked on with you. Did the candidate have trouble finishing the work?
6. What issues did you have with the candidate's work?
7. If I were the candidate's manager, what advice would you have for me?
8. How quickly did the candidate learn about the product or the product line?
9. How quickly did the candidate integrate with the rest of the team?
10. Why will or did the candidate leave?
11. Would you rehire or work with this candidate again?
12. What is the candidate's current or most recent salary?

Additional questions to ask when the candidate is to be considered for a management position:

13. How many people reported to this candidate?
14. How did the candidate conduct performance appraisals?
15. How did the candidate handle "difficult" people?

If you have questions that came out of the group interviews, now is the time to ask. For example: "One reservation some of our interviewers had was that the candidate seemed very dogmatic about certain process issues. We tend to decide about our project practices by consensus here. How do you think the candidate would work in this kind of environment, based on your experience?"

Template B-8: [Job Title] Reference-Check Form & Script.

Offer Letter Template

<[date]>

<[candidate name]>
<[candidate street address]>
<[candidate city, state, and zip code address]>

Dear <[candidate name]>:

I am pleased to offer you the position of <[job title]> *with* <[company name]>, *located at* <[location description]>, *reporting to* <[manager name and title]>.

    1. Your responsibilities will be those outlined in the enclosed job description and described to you during your discussions with me.

    2. You will be compensated with a <[weekly/biweekly/monthly]> *salary in the amount of* <[the salary]>. *Other compensation shall consist of* <[list of additional benefits/stock/perks]>.

    3. The Company has the following <[number]> *pre-employment requirements:* <[physical examination/review of documents]>, *which will need to be satisfied prior to employment.*

    4. You are considered an "at will" employee. This means that we can terminate your employment with or without cause, and with or without notice, at any time, at the option of either <[company name]> *or yourself, except as otherwise provided by law. Additionally, because you do not have an employment contract with us, you can terminate your employment with or without notice at any time.*

    5. Our offer to hire you is contingent upon your submission of satisfactory proof of your identity and your legal authorization to work in <[country name]>. *If you fail to submit this proof,* <[federal/state/local]> *law prohibits us from hiring you.* (Check whether you need this clause if you work outside the United States.)

    6. I hope you can begin work on <[day, date, time]> *at* <[position location>].

If you agree with and accept the terms of this offer of employment, please sign below and return this letter to our office on or before <[day, date]>. *I look forward to hearing from you and to having you join us.*

Sincerely,

<[your name]>   <[candidate name (signature)]>   *Date signed:* <[date]>
<[your title]>   <[candidate name (printed)]>

Template B-9: Offer Letter.

New Hire Orientation Checklist	
Activities to complete once the offer has been accepted	**Check off when task completes**
Order badge, keys, and key cards, as needed.	
Identify suitable office space, and verify that the space is clean and ready for a new occupant.	
Verify that the chosen office is equipped with a desk, lamp, chair, phone, and all necessary computer equipment, and that all are in working order.	
Order any needed furniture, office supplies, or computer equipment missing from the chosen office.	
Requisition an e-mail address, a voice-mail connection, and a physical mailbox.	
Activities to complete in preparation for Day One	
Stock the office with basic office supplies, such as pens, paper, pencils, wastebasket, scissors, stapler, staples, staple remover, and so on.	
Verify e-mail access and network computer hookup.	
Verify that phone and e-mail directories and location maps are available; add the new hire's voice-mail extension to the phone list.	
Identify locations for all applications and templates required by the new hire.	
Supply the new hire with printed manuals, as applicable to his or her work, or provide instructions for electronic access.	
Assign a buddy who can be available for the first month or so to answer the new hire's technical questions about how the team works and non-technical questions about staff, neighborhood, rules, and traditions peculiar to the specific environment and culture.	
Prepare a welcome letter and orientation package, including all HR forms.	
Activities to complete when the new hire arrives on Day One	
Add the new hire's name and title to the organization chart; add his or her name and extension number to the phone directory and other relevant lists.	
Introduce the new hire to project members, executives, personnel, administrative staff, as needed.	
Show the new hire instructions for calling meetings, for booking a conference room, and for other administrative procedures, as needed.	
Paperwork for the new hire to fill out on Day One, to be packaged with a welcome letter and orientation packet	
IRS, INS, and immigration forms, as applicable	
Health, dental, and life insurance forms, as applicable	
Benefits forms (long-term disability, short-term disability, and pension or retirement plans) as applicable	
Nondisclosure agreement, as applicable	
Emergency contact form	
Direct-deposit and check-cashing forms, as applicable	
Business cards, as applicable	

(continued on next page)

(continued from previous page)

Paperwork to give to the new hire to keep	
Maps, floor plans, and directions	
Parking, public transportation, and commutation information	
Personnel and HR policies (conflict of interest policies; sickness, holiday, and vacation policies; lateness and absence policies; sexual harassment policies; conflict of interest policies; medical and personal leave policies; birthday lists), as applicable	

Template B-10: Orientation Checklist.

The purpose of this book has been to make the hiring of technical people an easier, more enjoyable, and essentially manageable undertaking. The templates conclude the effort. Whether or not I have succeeded to the degree I hope, I welcome your comments and feedback. Please e-mail me directly—jr@jrothman.com—or contact me through Dorset House Publishing—rothmaninfo@dorsethouse.com.   I look forward to hearing from you.

Bibliography

Adler, Lou. *Hire with Your Head: Using Power Hiring to Build Great Companies*, 2nd ed. Hoboken, N.J.: John Wiley & Sons, 2002.

Buckingham, Marcus, and Curt Coffman. *First, Break All the Rules: What the World's Greatest Managers Do Differently*. New York: Simon & Schuster, 1999.

Cadwell, Charles M. *New Employee Orientation: A Practical Guide for Supervisors*. Menlo Park, Calif.: Crisp Publications, 1988.

Chowdhury, S. *The Talent Era: Achieving a High Return on Talent*. Upper Saddle River, N.J.: Prentice-Hall, 2002.

Cooper, Alan, and Robert Reiman. *About Face: The Essentials of Interaction Design*. Indianapolis: Wiley Publishing, 2003.

DeLuca, M.J. *Best Answers to 201 Most Frequently Asked Interview Questions*. New York: McGraw-Hill, 1997.

DeMarco, Tom. *Slack: Getting Past Burnout, Busywork, and the Myth of Total Efficiency*. New York: Broadway Books, 2001. (Distributed by Dorset House Publishing. ISBN: 0-932633-61-7)

———, and Timothy Lister. *Peopleware: Productive Projects and Teams*, 2nd ed. New York: Dorset House Publishing, 1999.

Drucker, Peter F. *The Essential Drucker*. New York: HarperCollins, 2001.

———. *Management Challenges for the 21st Century*. New York: Harper-Collins, 1999.

———. *The Practice of Management*. New York: HarperCollins, 1993.

Falcone, Paul. *The Hiring and Firing Question and Answer Book*. New York: AMACOM, 2002.

———. *96 Great Interview Questions to Ask Before You Hire*. New York: AMACOM, 1997.

Frame, J. Davidson. *Project Management Competence: Building Key Skills for Individuals, Teams, and Organizations*. San Francisco: Jossey-Bass, 1999.

Gause, Donald C., and Gerald M. Weinberg. *Exploring Requirements: Quality Before Design*. New York: Dorset House Publishing, 1989.

Hacker, Carol A. *The Costs of Bad Hiring Decisions & How to Avoid Them*, 2nd ed. Boca Raton, Fla.: CRC Press, 1999.

Harris, Jim, and Joan Brannick. *Finding & Keeping Great Employees*. New York: AMA Publications, 1999.

Janz, Tom, Lowell Hellervik, and David C. Gilmore. *Behavior Description Interviewing*. Englewood Cliffs, N.J.: Prentice-Hall, 1986.

Kador, John. *The Manager's Book of Questions: 751 Great Interview Questions for Hiring the Best Person*. New York: McGraw-Hill, 1997.

———. *201 Best Questions to Ask on Your Interview*. New York: McGraw-Hill, 2002.

Kaner, Cem, James Bach, and Bret Pettichord. *Lessons Learned in Software Testing: A Context-Driven Approach*. New York: John Wiley & Sons, 2002.

Lucht, J. *Executive Job-Changing Workbook*. New York: Viceroy Press, 2002.

———. *Rites of Passage at $100,000 to $1 Million+: Your Insider's Lifetime Guide to Executive Job-Changing and Faster Career Progress in the 21st Century.* New York: Viceroy Press, 2001.

Magretta, Joan, with the collaboration of Nan Stone. *What Management Is: How It Works and Why It's Everyone's Business.* New York: The Free Press, 2002.

Maister, David H. *Practice What You Preach: What Managers Must Do to Create a High Achievement Culture.* New York: The Free Press, 2001.

McGovern, Gerry, Rob Norton, and Catherine O'Dowd. *The Web Content Style Guide: An Essential Reference for Online Writers, Editors, and Managers.* London: Pearson Education Ltd., 2002.

Michaels, Ed, Helen Handfield-Jones, and Beth Axelrod. *The War for Talent.* Boston: Harvard Business School Press, 2001.

Mongan, J., and N. Suojanen. *Programming Interviews Exposed: Secrets to Landing Your Next Job.* New York: John Wiley & Sons, 2000.

Mornell, Pierre. *45 Effective Ways for Hiring Smart: How to Predict Winners & Losers in the Incredibly Expensive People-Reading Game.* Berkeley, Calif.: Ten Speed Press, 1998.

Reynolds, S. *Thoughts of Chairman Buffett.* New York: HarperCollins, 1998.

Rosse, Joseph, and Robert Levin. *High-Impact Hiring: A Comprehensive Guide to Performance-Based Hiring.* San Francisco: Jossey-Bass, 1997.

Ryan, Liz. "HR Interviews Can Be Senior Execs' Waterloo," *The Wall Street Journal's* Executive Career Site. *CareerJournal.com.*

Stanfield, B.R. *The Art of Focused Conversation: 100 Ways to Access Group Wisdom in the Workplace.* Toronto: New Society Publishers, 2000.

Still, Del J. *High Impact Hiring: How to Interview and Select Outstanding Employees.* Dana Point, Calif.: Management Development Systems, 1997.

Taguchi, Sherrie Gong. *Hiring the Best and the Brightest: A Roadmap to MBA Recruiting.* New York: AMACOM, 2002.

Weinberg, Gerald M. *Becoming a Technical Leader: An Organic Problem-Solving Approach.* New York: Dorset House Publishing, 1986.

———. *The Psychology of Computer Programming: Silver Anniversary Edition.* New York: Dorset House Publishing, 1998.

———. *Quality Software Management: Volume 1, Systems Thinking.* New York: Dorset House Publishing, 1992.

———, James Bach, and Naomi Karten, eds. *Amplifying Your Effectiveness: Collected Essays.* New York: Dorset House Publishing, 2000.

Wendover, Robert W. *Smart Hiring: The Complete Guide to Finding and Hiring the Best Employees*, 2nd ed. Naperville, Ill.: Sourcebooks, 1998.

Wilson, R.F. *Conducting Better Job Interviews.* Hauppauge, N.Y.: Barron's Educational Series, 1997.

Wysocki, Robert K., Robert Beck, Jr., and David B. Crane. *Effective Project Management*, 2nd ed. New York: John Wiley & Sons, 2000.

Yate, Martin. *Hiring the Best: A Manager's Guide to Effective Interviewing.* Holbrook, Mass.: Adams Media, 1994.

Index